CW01639503

The

The Story of the Fi
(Originally entitled

'one of the best-ever books on the Korean War . . . as vicious, uncomfortable, bloody and wasteful as any war in history.'

Manchester Evening News

'His account of the epic stand of the 1st Gloucesters on the Imjin River is magnificent . . . a vivid story of defiance against enormous odds.'

The Gloucester Citizen

'For anyone who would wish to know what the infantryman must suffer in battle there is no better account than this.'

Warwick Advertiser

'The story of the slow to anger giant, Bill Speakman, who earned his VC carrying on a one-man war with grenades . . . and 2nd Lieutenant William Purves, also a KOSB, who made history by being the only National Service officer to be awarded the DSO.'

Aberdeen Evening Express

'An eminently readable account of what will probably be the last occasion on which the nations of the Old Commonwealth will ever fight side by side in a conventional shooting war.'

The Times Literary Supplement

Also by the same author in Pan Books

THE FALL OF HONG KONG

CONDITIONS OF SALE

This book shall not by way of trade or otherwise, be lent, re-sold, hired out or otherwise circulated without the publisher's prior consent in any form of binding or cover other than that in which it is published and without a similar condition including this condition being imposed on the subsequent purchaser. The book is published at a nett price, and is supplied subject to the Publishers Association Standard Conditions of Sale registered under the Restrictive Trade Practices Act, 1956.

TIM CAREW

The Korean War

The Story of the Fighting Commonwealth Regiments

Originally entitled
Korea – The Commonwealth at War

UNABRIDGED

PAN BOOKS LTD · LONDON

First published 1967 as *Korea – The Commonwealth at War* by Cassell & Co Ltd.
This edition published 1970 by Pan Books Ltd,
33 Tothill Street, London, S.W.1

330 02474 4

© Tim Carew 1967

Printed in Great Britain by
Cox & Wyman Ltd,
London, Reading and Fakenham

Contents

Illustrations in Photogravure

All illustrations are reproduced by courtesy of the Imperial War Museum, with the exception of the one of Major Kenny Muir, VC

Hill 235 where Lt-Col Carne (Gloucesters) ordered his bugler blow the long reveille defying attackers in epic last stand

Vigilant Australian medium-machine-gun teams scanning hills for Communist activity

The Hook, behind the line in Korea, saw more enemy activity than any other Commonwealth front line position

Drum-Major P. Buss, of the Gloucesters' corps of drums, in the second batch of prisoners-of-war leaving for the UK on *Empire Orwell.* During the Imjin Battle the Drum-Major formed and commanded a platoon which took a heavy toll of the onrushing Chinese hordes

Battle of the Hook: Private Bill Morrison (20), Private Jim Walker (22), Private Jim Flynn (19), Private Tom Stanley (20), and Piper D. Ramsey

Maps

Acknowledgements

THIS is not an official history, very far from it – indeed, it is not a history of any sort. Neither is it an *exposé*, laden with revelations of military ineptitude at all levels.

It is simply the story – the long and true story – of a war and the men who fought it: a war as vicious, sordid, uncomfortable, bloody and wasteful as any in history.

It is primarily concerned with the soldiers of many different nationalities who came to the aid of South Korea and the United States in a conflict which was the responsibility of every right-thinking and free person in the world. It was a war which was won by neither side, but one victory *was* scored – a victory over Communism and all its vile tyranny.

In the Second World War and for seven years after it I served in an odd miscellany of units: I was a bewildered trooper in the Royal Horse Guards, a melodramatic corporal in the Parachute Regiment, a sort of John Masters-type subaltern of Gurkha Rifles, and a stern-faced captain and adjutant of the Devons. I was also a General's ADC for a period of three months (considering my shortcomings in the social world this represented a generous tenure of office), I endeavoured to instil officer-like qualities into cadets at OCTUs and occupied a few staff appointments, none of which called for an extended military education at Camberley.

My soldiering took me to India, Burma, the Netherlands East Indies, Hong Kong, Malaya, Eritrea, and Egypt. It would be nice to say that I served in Korea, but I did not. In July 1950, when Brigadier Aubrey Coad took his 'Woolworth Brigade' to war, I was unmartially employed counting soldiers' socks and subscribing to military miscalculation at the pay table in Bordon; in April 1951, as the Gloucesters formed up for the last fight on the Imjin, I was sweltering ineffectually in the Canal Zone of Egypt;

throughout the static war of 1952 I alternated between sticking pins into a map at Hounslow, inculcating fighting spirit into 'Z' Reservists at Larkhill and advising 'Cubby' Broccoli and Terence Young – my advice went largely unheeded – on the film *The Red Beret*; while the Black Watch and the Duke of Wellington's Regiment fought on The Hook, I was striving, in the main unsuccessfully, to master the writer's craft.

To make proper acknowledgement to all the people who assisted me in the writing of this book would call for a document similar in size and content to a Classified Army List. They range from General Sir James Cassels, GCB, KBE, DSO, who received me in an elegant but essentially functional office in Whitehall, to a Puerto Rican ex-private fortuitously discovered in a licensed betting office.

In addition to Jim Cassels, I met a number of other generals who gave generously of their time to help me: Major-General Aubrey Coad, CB, CBE, DSO, who commanded 27 Brigade at the very beginning; Major-General Tom Brodie, CB, CBE, DSO, who barred the way to the Chinese on the Imjin; Lieutenant-General Sir William Pike, KCB, CBE, DSO, then chief gunner of the British Commonwealth Division; Major-General John MacDonald, CB, DSO, OBE, who successively commanded the 1st King's Own Scottish Borderers and 28 Commonwealth Brigade; Lieutenant-General Sir Geoffrey Evans, KBE, CB, DSO, who prepared the soldiers in Hong Kong for the Korean War; Major-General John Sloane, CBE, who succeeded the late Major Kenny Muir, VC, as second-in-command of the 1st Argyll and Sutherland Highlanders; Major-General Abdy Ricketts, CBE, DSO, commander of the 29th British Brigade in 1952.

A galaxy of brigadiers, colonels and majors swam into my ken as I scoured the British Isles for men who had fought in Korea: Brigadier Anthony Farrar-Hockley, DSO, MBE, MC (Gloucesters), Brigadier James Majury, MBE (Royal Ulster Rifles), the late Brigadier George Butler, DSO, MC (8th Hussars), Brigadier Dennis Tadman, DSO, OBE (King's Own Scottish Borderers), Colonel Andrew Man, DSO, OBE (Middlesex), Colonel Denis Harding, DSO (Gloucesters), Colonel Aymar Clayton, OBE (Middlesex), Lieutenant-Colonel John Deighton, OBE, MC

(Royal Northumberland Fusiliers), Lieutenant-Colonel Denis O'Flaherty, DSO (Royal Artillery), Lieutenant-Colonel Jim Shelton, MC (3rd Battalion Royal Australian Regiment), Lieutenant-Colonel Sandy Boswell, MBE (Argyll and Sutherland Highlanders), Lieutenant-Colonel Christopher Troughton, MC (8th Hussars), Lieutenant-Colonel Hugh Hamill (Royal Ulster Rifles), Major John Winn, DSO, MC (Royal Northumberland Fusiliers), Major 'Spike' Pike (Gloucesters). I have a special word of thanks for Brigadier Neville Grazebrook, CBE, DSO, who made all the records of the Gloucestershire Regiment available to me and Colonel Bill Ponting, OBE, who was always on hand with suggestions for new contacts, as was Major Brian Holdsworth.

Many others helped me from a distance. I have not met them personally, but hope to rectify this some day: Lieutenant-Colonel Edgar Bell (United States Army), Lieutenant-Colonel Reggie Pratt, DSO (Royal Northumberland Fusiliers), Lieutenant-Colonel Tom Slessor (Argyll and Sutherland Highlanders), Lieutenant-Colonel Philip Harrison, DSO, OBE (King's Own Scottish Borderers), Lieutenant-Colonel Humphrey Wilson, MC (Royal Norfolk Regiment), Major Duncan Bengough (Black Watch).

The term 'a subalterns' war' requires some qualification: every war is a subalterns' war because the platoon commander is always the man who sees most of the game. Of the many Korea subalterns I met, some are still serving and have become majors and lieutenant-colonels with a hint of grey at the temples and impressive rows of medal ribbons earned in post Second World War campaigns – John Moncrieff (Black Watch), Sandy Boswell, Christopher Troughton, Mervyn McCord, MC (Royal Ulster Rifles), Christopher Lawrence, MC (Middlesex), Alan Hill (Royal Ulster Rifles), Paddy Baxter (Royal Northumberland Fusiliers).

Others have left the Army and occupy various placings in the civilian rat-race of 1967; Stanley Fothergill (Middlesex), Hugh Sawbridge (Durham Light Infantry), Keith Taylor (Royal Northumberland Fusiliers), Denis Clough (King's Shropshire Light Infantry).

The writing of this book would have been virtually impossible without the constant help of Lieutenant-Colonel Walter Young, MBE, Librarian at the Staff College, and Mr Derrick King and his assistants in the Ministry of Defence Library.

Finally, the 'other ranks' – there seems to be some doubt as to their correct nomenclature, so I will call them soldiers because that is the term that I suspect they would like best. I found them in all manner of improbable places, pursuing a wide variety of occupations: driving lorries, clerking, keeping pubs, labouring.

I was fortunate enough to find four of the original Korea battalions stationed in the United Kingdom: the 1st Royal Ulster Rifles at Bulford, 1st King's Own Scottish Borderers at Shorncliffe, 1st Royal Northumberland Fusiliers at Kirton Lindsey, and the 1st King's Shropshire Light Infantry at Crownhill, and all four were generously leavened with Korea veterans. Lieutenant-Colonel Hugh Hamill, Lieutenant-Colonel Colin Stonor, MC, Major Paddy Baxter and Major Bernard Houghton-Berry extended to me the type of VIP treatment and hospitality normally reserved for Cabinet Ministers. Warrant officers, sergeants, corporals, lance-corporals and privates were delivered to me with broad grins, a stimulating crash of heels and a wealth of invaluable information which will go a long way to enhance any success that may attend this book.

Tim Carew

PART ONE

Without Benefit of the British Commonwealth

JUNE–SEPTEMBER 1950

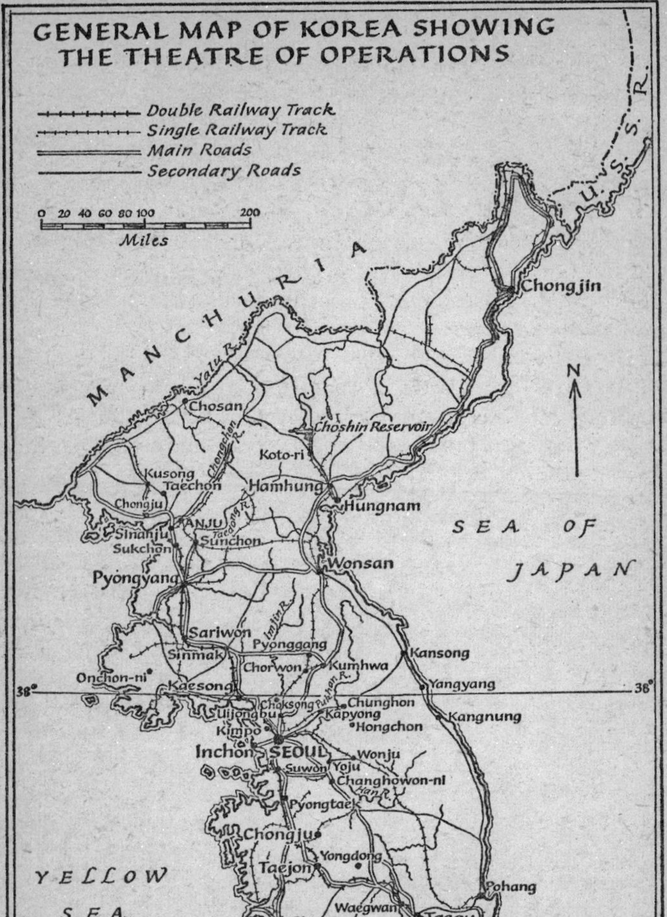

© CASSELL & CO LTD 1967

Prologue

THE ribald phrase 'they were caught with their trousers down' has been part of the English language for a little over a quarter of a century: it has been used freely in conversation to illustrate any embarrassment ranging from an industrial tycoon caught by the tax collector to a country suddenly and wantonly invaded by an enemy.

The phrase has certainly taken its place in the history of war, and so far as Great Britain is concerned it probably started with the Vikings, who caught our island in a condition of *flagrante delicto* from which it would appear that she has never fully recovered.

As the Second World War gathered momentum, the phrase was used with increasingly deadly effect: Dunkirk, Hong Kong, Singapore, and Burma. A platoon in Burma was caught with its trousers down (some of its members literally) because they were brewing tea when a marauding party of Japs hit them; the British Expeditionary Force in France were, man for man, in a similar predicament when the Germans stormed into Belgium. The Abyssinians were caught by the Italians, the Finns by the Russians, the Russians by the Germans.

Probably no nation's pants were nearer ground level than those of the United States on December 7th, 1941, the day of infamy which is loosely known as Pearl Harbor. In the annals of unpreparedness Pearl Harbor must carry off any palm and no historian, however biased, has ever seriously argued about this.

The inelegant reference to trousers at ankle level must be used once more in relation to South Korea and the United States in 1950, but not, for once, the British Commonwealth which was sadly bereaved, continuously bewildered but securely buttoned up at waist level.

ONE

KOREA is best known to its own people as Chosen, 'land of the morning calm'. The western name Korea was derived from the name of the Koryo dynasty of AD 935 to 1392; literally translated it means 'land of high mountains and sparkling streams'. Except for the reference to high mountains, these grandiloquent descriptions scarcely matched the events between July 1950 and July 1953. To the soldiers of many nationalities who fought there it was a pestilential area of smells, all of them foul.

Extending from Manchuria and USSR, the peninsula of Korea is 525 miles long. At its narrowest it measures 125 miles, at its broadest 200 miles. Dotted haphazardly along its 5,400 mile coastline, like 'hundreds and thousands' sprinkled on the icing of a cake, are more than 3,000 small islands.

The mountains are everywhere, and soon they will have names: Pork Chop Hill, Big Nori, Old Baldy, Gloster Hill, Jane Russell, Betty Grable, The Hook, The Brown Bastard – but that time is not yet. In June 1950 they are just a profusion of hills – towering, rugged, menacing, inhospitable.

Geographical statistics will tell you that the climate of Korea is humid and continental, that it has a pronounced rainy season and relatively dry winters, that temperatures vary from 105° F in summer to – 40° F in winter. What they will not tell you about is the blinding heat, the choking dust, the torrential rain and the cold which turns the peninsula into a frozen wilderness.

The 38th Parallel, which was to achieve such significance in the Korean War was, in fact, nothing more or less than an imaginary line and artificial barrier between north and south; a convenient way of dividing Korea into two sections following the end of the war with Japan.

The Japanese soldiers to the south of this line surrendered to the United States, those to the north to the Russians. But it was by no means as simple as that: the 38th Parallel in effect became the iron curtain of Korea.

It sealed off north from south with peremptory thoroughness, paying no regard to the political inclinations of the inhabitants, if indeed they had any: the 38th Parallel cut through provinces, counties and towns and through rivers, mountain ranges, and railway lines. Of Korea's thirty million people, there were now nine million in the north and twenty-one million in the south.

The north was industrial and possessed most of the country's gold, wood, iron, copper, and graphite and nearly all of its hydro-electric power. All Korea's riches were in the north; the south was exclusively agricultural, and for the peasants with their primitive farming methods life was an all-the-year-round struggle for existence barely above the poverty line.

[2]

In 1950 America was sick of the mention of a war which had decimated her young male population in Europe and the Pacific. Great Britain and the Commonwealth had had more than enough, but Britain's hands were still militarily full: there were belligerent Yugoslavs in Trieste, bandits in Malaya, verminous miscreants in Eritrea, and prickly Egyptians – all perversely intent on delaying a full return to peace for as long as possible. Britain, tired and exhausted as she was by six years of war, was still playing her traditional role as a global policeman.

Soviet Russia, however, was just beginning to enjoy herself. As far as post-war troubles were concerned, the USSR resembled the guest at a party who gets drunk and in spite of the broadest of hints refuses to leave; to use a widely used contemporary phrase she had been 'stirring it up' since the end of the Second World War. By subverting national elections the nations of central Europe had been put under the control of puppet governments. Russia had become an atomic power, sooner than had been expected, and told the world about it with truculent authority. Communist guerillas started trouble in Greece, and she and Turkey all but disappeared behind the Iron Curtain – only the Truman Doctrine of 1947 prevented this.

In 1948 Russia established a land blockade against Western

Berlin, and a gradually awakening America helped by an exasperated Great Britain flew 2,300,000 tons of food and coal into the city between April 1st, 1948 and September 30th, 1949.

But it was in the Far East that the Communist menace was most apparent. Red terrorism seemed to be gaining ground in Malaya; Mao Tse-tung defeated Chiang Kai-shek in China; there were 100,000 Russian soldiers in North Korea, and other Soviet troops were systematically overrunning Manchuria.

[3]

In June 1950 the armies of South and North Korea peered warily at each other across the 38th Parallel.

The soldiers of the two armies resembled one another in looks but in nothing else: the force below the Parallel was no match for the one above it, either in numbers, arms or experience. The South Koreans were about as ready for war as the United States had been at Pearl Harbor or Great Britain at Singapore.

On paper the South Korean Army looked impressive enough: it mustered 65,000 men armed with weapons used by American soldiers and marines in the Second World War: M1 rifles, carbines, light and heavy machine guns, mortars, and bazookas. But they had been conditioned by benign American occupation, and they had no tanks or air force.

The North Koreans had tanks and aeroplanes, in addition to 90,000 men trained by the infinitely more persuasive methods of the Russian 'advisers'.

The South Koreans were outnumbered and outgunned – the biggest South Korean artillery pieces had a maximum range of 8,200 yards against the 14,000 yards of the guns in the North. The North, geared for war, had enormous reserves of ammunition and a vast fleet of fast and well-maintained vehicles; the South, seemingly prepared only for casual summer manoeuvres, had only enough ammunition for a few days and almost half their transport was unserviceable.

The officers of the North Korean Army were tough, dedicated professionals: many of them had fought with the Communist

Chinese forces, and many others had received intensive and specialized training in the Soviet Union.

The South Korean officers were also tough, for physical hardiness, for any Korean, is a matter of grim necessity in the business of living, whatever his politics. Few of them, however, were either dedicated or professional – many battalions and regiments were led by young, inexperienced colonels and even generals in the twenty-five to thirty age group.

In Korea the night of June 24th, 1950 was a foul one. It was the beginning of the summer monsoon, a three-month period of torrential downpours. On that night the rains were heavy and unrelenting and the South Korean soldiers manning the outposts along the 38th Parallel were taking a soaking. They accepted the driving rain equably because they were an agricultural nation and the rain would guarantee the rice and barley crops which would keep them alive.

Just before the dawn on June 25th, 1950, the sodden soldiers on the slopes of the 38th Parallel heard what they thought was thunder and saw flashes in the sky which they took to be lightning. But the rumbling noise came from the artillery of the North Koreans and the flashes in the sky heralded the shells which were soon crashing into their positions.

Thus started the Korean War . . .

[4]

As so often happens in war, the best were the first to die: so it was when the North Koreans launched a two-division attack with strong tank support on the Pyongyang-Seoul highway. The South Koreans fought back valiantly, but they were never seen with a chance – the American anti-tank guns and bazookas were little better than catapults against the Russian T-34 tanks. Squads of volunteers, displaying either extraordinary heroism or extraordinary stupidity, depending on one's conception of war, climbed on to the turrets of the tanks and tried to drop hand grenades inside. They were probably as brave as any soldiers in any war before or since, but they did little to stop the steamroller advance: a few blew themselves to pieces on their own grenades; others – not a few – were contemptuously swept off the turrets by machine-gun fire or picked off by riflemen. And soon, as Robert Leckie neatly phrased it in his book *The Korean War*, 'there were no more volunteers for suicide duty'.

This early stubborn resistance was not maintained; perhaps it is unfair to expect that it could be. By the evening of June 25th, Kaesong was in North Korean hands and the straight road south to Seoul lay open to the invaders.

On the second day of the war the South Korean Army began to fall apart as a cohesive military unit, although some of the less charitable military experts question whether it had ever been one in the first place. In the same breath they heap criticism on the United States for not building up military strength in the South on the same scale as had the Russians in the North.

On this day of crushing defeat the deplorable staff work of the South Korean Army was mercilessly exposed: heavy guns and vehicles were abandoned; regiments unashamedly broke and ran, or stood their ground and were annihilated where they stood, or launched into despairing and unplanned counter-attacks, which were mown down like stubble.

Road blocks were abandoned and natural defensive features were left uncovered. Isolated pockets of brave men either failed to receive, or ignored, orders to withdraw. There were units who, when ordered to withdraw, plunged desperately on towards the enemy and other units, detailed to counter-attack, which retreated in ignominious confusion.

Bridges carefully primed for demolition were not blown and orders were given to blow bridges which had not been primed. Colonels were relieved or relieved themselves of their commands. Generals fumed and fulminated, cursed, wept and committed suicide, carried out one-man attacks on tanks, and fled from the stricken field.

The American Advisory Group, shouldering a burden which was rapidly becoming intolerable, did their best, although it did not amount to much: their advice was usually either ignored, misunderstood or went unheard. In the words of John Dille, a tough and seasoned American newspaperman, the American Advisory Group resembled Frankenstein and the South Korean Army his monster – they had created it but could not control it.

The roads, poor at the best of times, were clogged with transport and leaderless soldiers. Vehicles crawled bumper to bumper; the soldiers, many of them wounded, struggled on with aimlessly lagging footsteps. And many of these wounded would be dead before nightfall, for gangrene strikes swiftly in a country where crops are fertilized with human excreta.

To add to the overall confusion there was another menace, the most beastly and tragic aspect of war: the civilian refugees. The roads and the bridges across the Han River were black with humanity: old men and women, wearing the classic expressions of Oriental inscrutability and carrying all their worldly possessions in hastily wrapped parcels and on wheelbarrows; young women, whose husbands were dead, dying, wounded, still fighting or endlessly marching, had babies on their backs or at the breast; children too old to be carried and too young to fight; babies abandoned to die because the rest of the family could carry no more.

Occasionally, stoicism gave way to animal fear and despair and a piteous wail would start somewhere along the column, to be taken up by others who still had breath left for crying. But mostly they trudged southwards in weary silence on roads which seemed to have no ending. They had their lives and that was about all.

Close on a thousand of them were to lose their lives on the bridges which crossed the Han River. By one of those appalling errors of judgement which had become almost commonplace in the South Korean Army, the bridges were blown at a time when the human traffic crossing them was at its thickest. This was tragedy enough, but there was worse to come: the greater part of the army was still north of the river when the bridges crumbled. Wantonly beleaguered on the wrong side of the Han River, the remnants of the South Korean Army were now in the direct path of the advancing forces from the North.

Russia took full advantage of the situation through the dubious medium of the Communist-controlled radio station at Pyongyang.

A glib-tongued announcer told of an attempt by the 'bandit traitor Syngman Rhee' to invade North Korea, and with gathering malevolence went on to say that the 'righteous invasion' of Communist forces had been launched only as a counter-measure.

By the end of June 1950, the South Korean Army had been reduced to 22,000 men, although for the next fortnight stragglers continued to wander in. The term 'straggler' has always had a contemptuous ring about it, but few South Korean stragglers were guilty of any dereliction: they kept going until they were killed, captured or finally reached their own lines – some with no helmets, boots or arms, some with shirts but no trousers, some with trousers but no shirts; some with neither.

The melancholy figures of killed, wounded and missing spoke for themselves, even allowing for the hit-or-miss methods of the South Korean Army in reporting casualties – 44,000 men had been lost in addition to 30 per cent of the army's small arms and almost all its transport and artillery. It was defeat, complete and utter, for South Korea and, indirectly, for the United States.

On the afternoon of June 28th, Seoul fell to the North Koreans, and the United States found that they had another war on their hands.

TWO

IF America was not exactly caught with her pants down, they were at best at half-mast.

On Saturday June 24th, 1950, at 9.26 PM Washington time, John J. Muccio, United States Ambassador in Korea, informed the State Department that the Communist attack appeared to constitute an all-out offensive against the Republic of Korea.

Secretary of State Dean Acheson, relaxing on his farm at Sandy Springs, Maryland, immediately telephoned President Truman who was at his home in Independence, Missouri. At 3 AM on Sunday morning Ernest A. Goss, deputy United States representative to the United Nations, telephoned Trygve Lie at his Long Island home. On receipt of the news Trygve Lie, a man not normally given to melodramatic utterances, declared: 'This is the beginning of World War Three.'

The United States as a whole took a much less dramatic view of this threat to a hard-won world peace. The immediate reaction in American homes from Manhattan to Monterey was one of disinterested incredulity – where was Korea anyway?

But Harry Truman was taking the Korean War seriously and put a swift damper on national complacency. 'The Communists in Korea,' he said, 'were acting just as Hitler, Mussolini, and Japan had acted ten, fifteen and twenty years earlier. If South Korea was allowed to fall, Communist leaders would be emboldened to override nations closer to our own shores. And if this was allowed to go unchallenged, it would mean a Third World War.'

At 2.20 PM on June 25th, as the first soldiers in the Korean War were dying, the Security Council of the United Nations assembled to discuss the invasion of South Korea in the light of a threat to world peace. By 6 PM the United States resolution calling for a cease-fire and North Korean withdrawal to the 38th Parallel had been passed.

The North Korean refusal to pay any heed to this resolution came as no surprise to anyone, and on June 27th the vote was taken on the American resolution that members of the United Nations 'furnish such assistance to the Republic of Korea as may be necessary to repel the armed attack and to restore the international peace and security in the area'. Britain, France, Nationalist China, Cuba, Ecuador, and Norway voted for the resolution; Egypt and India abstained, and Communist Yugoslavia voted against it.

[2]

While forward elements of the North Korean Army were nosing into the outskirts of Seoul, General of the Army Douglas MacArthur was taking a look for himself. He flew to Korea and drove north as far as the Han River, beyond which the North Koreans were regrouping for the rush which would overrun the entire peninsula. It was MacArthur's sombre report of the situation that finally put the United States into the war:

> The only assurance for holding the present line and to regain later the lost ground is through the introduction of United States ground combat forces into the Korean battle area . . . if authorized it is my intention immediately to move a United States regimental combat team . . . and to provide for a possible build-up to a two-division strength from the troops in Japan for an early counter-offensive.

This was it: there was now, clearly, no possible excuse for withholding United Nations aid to South Korea any longer.

But, on second thoughts, America decided that perhaps it wouldn't be so tough after all: there would not be another Bataan, Omaha Beach or Okinawa. It would be some sort of police action, a simple matter of showing the Stars and Stripes to a bunch of ragged-assed Commies who'd got a bit above themselves.

The North Koreans, however, were about to show them how wrong they were: Communist strategy required a quick end to the war in order to establish a Communist Korea before the United States and the United Nations could arrive on the scene in force.

The race for Pusan had begun.

The invasion of South Korea caught the United States Army of occupation in Japan with its pants down – in many cases literally.

In Japan the cherry blossom lay like delicate snow on the lush green countryside.

General Douglas MacArthur had done a good job on Japan – possibly too good: under his benign rule Japan's military machine had been destroyed, religious freedom had been proclaimed, sweeping land reforms had been instituted, large and greedy industrial combines had been broken up, a new constitution which demoted the Emperor and denounced war had been drafted, and women had become emancipated. Never in the world's history have there been more benevolent conquerors than the Americans in Japan.

For the American soldiers there was a profusion of bars, beer-halls and dance-halls; there was boogie-woogie and baseball; Scotch and sixteen-ounce steaks. Corporals rented villas and teenage privates bought cars. Such military chores as marching, shining boots, and garbage detail were relics of a more austere past. An American soldier in Japan rarely walked anywhere – he drove or was driven. And, of course, there were girls to suit every taste – geisha girls, taxi girls, call girls; hostesses, professionals, and enthusiastic amateurs. The American GI had it made in every conceivable department.

'Some of these guys,' reported a top sergeant, 'are so goddam fat they can hardly walk. They've forgotten what it's like to put their own pants on – the geisha girls do it for them. If this man's army ever goes to war, there'll be four guys in the line and eight bringing up the Coca-Cola.'

It was a harsh assessment of military might, but perhaps the only possible one to come from a soldier of seventeen years' service who had sweated out a score of bloody encounters on Guadalcanal. For better or worse, these men were the 'United States ground combat forces' destined for the Korean War. It had to be them, because no others were immediately available – the nearest Marines, America's elite of fighting men, were in California.

Four divisions were in Japan under General MacArthur – the 7th, 24th and 25th Infantry, and the 1st Cavalry. A United States infantry division was somewhat different in composition to its British counterpart with its classic complement of three brigades of three battalions each.

Prior to 1939 the United States Army had been organized into 'square' divisions, so-called because their structure included two infantry brigades of two regiments each, with a total strength of approximately 20,000 men. At the beginning of the Second World War the expression 'brigade' disappeared with the formation of 'triangular' divisions consisting of three infantry regiments of three battalions each.

In Japan events in Korea had been regarded with casual

disinterest. If there was going to be war, then it would be a short and easy one. Indeed, it might not even be a war at all, and the fatal expression 'police action' was on everyone's lips.

The Japanese press was bland and non-committal. 'There is no need for hysteria among the Japanese,' said a cautious leader writer in the *Nippon Times*, 'but there should be a calm appraisal of the situation, for a civil war in Korea represents trouble which is too close for comfort.'

The stage was set for tragedy. But the cast of thousands was very far from being ready . . .

On the night of June 24th, as the North Koreans reached their assembly points above the 38th Parallel, Major-General William Dean was wearing the long robes of a Korean *yang-ban* (gentleman) – or, as Dean himself neatly phrased it, 'a guy who doesn't have to work'.

The occasion was a fancy dress ball at Kokura, on Kyushu, the most southerly Japanese island, for the officers of the 24th Division. Like the Duke of Richmond's ball on the eve of Waterloo, it was a splendid function: a major of signals, who had been in the bar longer than most people, did a hula-hula dance clad in a grass skirt, an artillery colonel, who had been there even longer, briefly officiated on the drums and broke one of them.

Dean was no stranger to war or to Korea: he had commanded the 44th Infantry Division in Europe in the Second World War and went to South Korea in 1947 as military governor. General Dean felt no ill effects the following morning; he was, in fact, as physically fit as many a soldier in Japan thirty years his junior, and considerably more so than most of his officers.

On the morning of the 25th Dean's thoughts were far from war; he was looking forward to collecting his mail. On his way to the post office he was approached by a harassed-looking staff officer who told him that the North Koreans had attacked the South.

And on receipt of this information Dean too said: 'This is the beginning of World War Three.'

[3]

At 4 AM on June 30th, Harry S. Truman launched the United States fairly and squarely into the Korean War with the single crisp order: 'Go ahead and send the troops.' He did not, however, use the phrase 'war': presumably to cover failure to ask Congress for a ratifying declaration of war, the President used the more guarded expression 'police action'.

Truman freely confesses that he gave this order with uneasy misgivings. 'I wanted to take every step necessary to push the North Koreans back behind the 38th Parallel,' he wrote in his memoirs, 'but I wanted to be sure that we would not become so deeply committed in Korea that we could not take care of such other situations as might develop.'

General MacArthur took a more soldierly view: he had seen at close hand the headlong retreat of disorganized South Korean troops and the refugee-choked roads – all the depressing ramifications of a defeated army. He had heard, also at close hand – too close for some generals – the sound of gunfire. American policemen, in the course of their duties, often have to shoot to kill. But they do not, so ran the unsubtle reasoning of Douglas MacArthur, use tanks and artillery. MacArthur did not use the expression 'police action' then or subsequently.

'To me it was war from the day I got there,' he said uncompromisingly, 'unadulterated war of the most savage character.'

To help stem the North Korean thrust towards Pusan the South Koreans needed the assistance of the British Brigade of Guards, the Gurkhas, the Foreign Legion, German SS men, Japanese Imperial Guards, the United States Marines who had held Wake Island or a force which was a synthesis of these, the world's most stubborn fighting men. They needed men who if they could not stay in a defensive position alive would stay in it dead. They needed men who were fighting for a cause, and believed in it.

They got none of these things. They got instead the United States 24th Infantry Division.

Major-General William Dean was a big man in every way – a

tough and dedicated soldier, even though he never went to West Point, and once worked as a stevedore, trolley conductor, and dishwasher.

Dean took the 24th Division to war with a list of material deficiencies which would have daunted Genghis Khan. Apart from its outdated Second World War complement of weapons, the division's vehicles were in a deplorable condition: some would not start when orders to mobilize were received; radiators were clogged, and overheating of motors was frequent; due to the pernicious aversion to marching of occupation troops in Japan, tyres were dangerously worn.

The 3rd Battalion of the 35th Infantry Regiment owned only one serviceable wireless set; the 24th Regiment confessed that it only had 60 per cent of its allowance of sets and that four-fifths of them were inoperable. The 1st Battalion of the 35th Infantry had no spare barrels for machine guns and the bipods and tubes of their mortars were worn out. There was a serious shortage of trip flares, cleaning rods, grenades, and anti-tank mines. Illuminating shells did everything except illuminate – when fired something like 60 per cent of them proved to be duds. It all added up to a logistical nightmare, but it was too late to do anything about it.

If the 24th Division was lacking in material matters, discipline and morale were in even shorter supply. The basic trouble was that many of the men were not really interested in being soldiers. The recruiting posters in the United States had promised world-wide travel to exotic places (Japan more than fulfilled this particular promise), generous pay, good food, comradeship, sport, and liberal furlough. They did not, however, suggest that the primary function of an army was to fight. Who was going to start another war, anyway?

'They were,' said a cynical newspaperman, 'just a bunch of kids – the kind of kids you send out to a movie so as you can get a bit of peace and quiet at home.'

This, then, was the 'arrogant display of strength' with which General MacArthur hoped to fool the enemy into the belief that he had much greater resources at his disposal than was actually the case.

So they went to war . . .

[4]

There was nothing in United States Army recruiting literature to describe a country like Korea. The men of the United States 24th Infantry Division disembarked at the port of Pusan and did not like what they saw and smelt.

This was understandable. For Pusan, although Korea's major port, is mostly a foul and sprawling slum. Behind the sheds, warehouses and derricks of the dock area lie the goods marshalling yards and stores depots. All this dock paraphernalia has grown in the midst of a sordid shanty town of flimsy and filthy wooden shacks which house most of Pusan's population.

Pusan – pronounced Poos'n by American soldiers – at first sight was bad enough; the smell was something else again. It is a smell which, to a greater or lesser degree, pervades the whole of Korea, a smell compounded of rice-paddy muck, mud walls, rotting garbage, and human excreta – particularly the latter. In Korea men, women and children are in the habit of performing their natural functions wherever they happen to be.

But in spite of its obvious defects, Pusan did its best for the disembarking soldiers. Flags, banners and streamers were everywhere, and brass bands blended in excruciating harmony with the excited shouts of the populace.

The optimists opined that it would be a short war and an easy one; the pessimists speculated morosely on the possibility of being taken prisoner and sent to the salt mines.

But they were all agreed on one thing: they didn't know what this war was all about, and they didn't care. Some of the officers had hinted vaguely that it was against the Communists, but in 1950 the word Communist did not have the same ring of menace to an American that it had so shortly afterwards.

Americans in the Second World War had gone to battle as crusaders, to fight for an impressive array of Many Splendoured Things – decency, civilization, democracy, freedom; they were fighting for Uncle Sam, mom and pop and the kids back home.

But the men of 24th Division arriving at Pusan did not see themselves in the role of crusaders. They had been sent at short notice from the fleshpots of Japan to some country which was all mountains and excreta. They did not like the idea and said so loudly and plaintively.

Marguerite Higgins, an American newspaperwoman who covered the Korean War as vigorously as any of her male colleagues and considerably more so than some of them, overheard a conversation between some GIs on the road to Taejon. This particular oration has been preserved as a classic statement of military non-intention:

> Somebody gave old Harry the wrong dope on this war. He can find someone else to pin his medals on. Just give me a jeep and I know which direction I'll go in. This momma's boy ain't cut out to be a hero.

The 24th United States Infantry Division added little lustre to American military history; perhaps it was unfair to expect that they should: the soldiers of the 24th proved no match for the hardy North Koreans who ran rings round them from the very start.

The North Koreans relentlessly developed frontal pressure by determined attacks by infantry assault groups, frequently supported by tanks. While the bewildered American defenders were pinned down, flanking forces moved with incredible speed to the rear: these flanking forces established road blocks and completed the envelopment of battalions, regiments and finally the whole of the 24th Division.

Since their first action on July 5th they had been outflanked, out-gunned and out-manoeuvred by an enemy superior in numbers, armament and, perhaps most important of all, discipline. The first three weeks of fighting represented practically unrelieved disaster for American ground forces.

At Osan, Pyongtaek, Chonui, on the Kum River Line and at Taejon the melancholy story of defeat and disorderly retreat was the same – confusion at all levels, inferior weapons, faulty communications, shortage of ammunition and, most damning of all, reluctance to fight.

Helmets and weapons were thrown away; some units 'bugged out' in the face of North Korean tanks while others sat down to await capture. Wounded were abandoned to the enemy and the fate of these unfortunates was a grim one: many of them were tied up and wantonly murdered with bullet and bayonet.

The North Koreans and the Korean climate relentlessly sought out the weak spots in the American soldiers. Marching had never figured prominently in their training in Japan and the hardy North Koreans, who could cover the mountainous country like antelopes, constantly outflanked and ambushed units riding in trucks.

Dysentery, that crippling disease which thrives on flies, bad sanitation and lax hygiene discipline, claimed victims which in some units amounted to as many as one in four; countless men succumbed to heatstroke, and salt tablets became top priority for supply air drops.

Map reading, like marching, had not received anything like the attention it deserved. Nor did the maps help much; they were based on a casual Japanese survey between 1918 and 1932. Roads turned out to be rivers, hill features had been unaccountably flattened into rice paddies. Motorized units – and the rueful observation was often made that no unit was truly motorized unless there was one jeep per man – struggled with contours and conventional signs and all the other mysteries inescapable from map reading, and drove straight into the enemy to be raked with machine-gun fire or, in many cases, to be sent into ignominious captivity. Artillery fire was inaccurate and shells were showered on the enemy, on friendly units or on nothing at all in equal profusion. Bazooka ammunition failed altogether to penetrate North Korean tanks.

Few soldiers fight well on an empty stomach, but the North Koreans did the next best thing – they fought and marched incalculable distances on three rice balls a day, a diet which made little difference to their health or their fighting efficiency. The Americans, deprived of a benevolent PX – the all-embracing canteen service – extracted what nourishment they could from Second World War K rations, which contrasted unfavourably with steaks, ham and eggs, ice cream, and mom's apple pie. A K ration consisted of a tin of compressed meat hash, coffee, powdered milk, a fruit bar, cigarettes, chewing gum, and toilet paper – the cigarettes, toilet paper, chewing gum were OK (although there was never enough of any of them), but the soldier's comments on the remaining commodities are not for print.

Of the many virtues of the American soldier, fire discipline has never been the most prominent: in any theatre of war their expenditure of ammunition has always been prodigious, and Korea in those early days was no exception. The reason for this was not hard to find. To green troops, in action for the first time the most frightening feature of battle is not the fear of death, dishonour or mutilation but *noise*.

Front-line war correspondents, and others less near the front line, have frequently referred to nights of confused and bitter fighting against 'hordes of enemy'. (Query, how many constitutes a horde?) In actual fact, the confusion lasted all night and the bitterness, like the proverbial refuse cart which follows the Lord Mayor's show, came next morning from harassed quartermasters.

A harsh example of the pernicious trigger-happy bug was to be found in the ranks of almost any infantry battalion in the early fighting. Men were jittery, hungry and sleepless. The blazing heat of the day had given way to alternating rain and mist. There were rumours of hordes of enemy to their immediate front.

Suddenly, for no attributable reason, a machine gun opened up in short stuttering bursts. It was the signal for all hell to be let loose: more machine guns opened up, rifles and hand grenades, as if determined to add their quota of foul din, joined in indiscriminately.

At length the murky and uneasy dawn brought with it a merciful quiet, and soldiers gingerly sallied forth from their foxholes to count the bodies which must be piled high all round the position.

There was one dead Korean outside the position. Judging by the smell, the body had been there for upwards of a month. The Americans counted not dead bodies but expended ammunition.

The 24th Division bought its battle experience at a fearful price.

In these first seventeen days of fighting two enemy divisions consisting mainly of infantry had driven it back 100 miles. It had suffered more than 30 per cent casualties and the figure of 2,400 missing was depressingly significant. It had also lost enough arms and equipment to equip another division.

Losses in officers of field grade were especially severe. During both world wars the attitude of some regulars towards their amateur comrades added to the unpleasantness of war, and the odious phrase 'temporary gentleman' was given full sway: apparently an education at Sandhurst, West Point or St Cyr imparted a soldierly gloss that lesser mortals could never attain.

But in the early days in Korea there was some justification for this feeling of professional superiority: in many cases, leadership among the officers had to be exceptional to persuade the enlisted men to fight, and many gave their lives in this effort. A large percentage of these were graduates of West Point, and it had been dinned into them that in war they would probably have to be the first to die. In Korea they were.

It is easy to be wise after the event and to condemn the performance of the 24th Division as undistinguished, but the hard facts are that units *did* 'bug-out', abandoning their wounded to the enemy; they *did* throw away arms, helmets, and equipment; they *did*, in some cases, sit down and await capture.

On the credit side, however, there were many individual acts of gallantry, and no officer or enlisted man behaved with more splendid disregard for his own safety than the Division's commander, Major-General William Dean.

From the start Dean knew the inadequacies of his troops – 'How in hell am I going to teach these boys they can't always jeep to battle?' was his most frequent lament. He set himself the task of remedying some of these defects by personal example and drove himself to physical and mental exhaustion in the process.

Dean broke all the accepted rules of war: he seemed to be everywhere at once, wherever the fighting was most savage: he joked, cajoled and encouraged; he congratulated the brave and heaped fearful derision on the shirkers; he excited unwilling admiration in the surliest and most recalcitrant GI.

He armed himself with a bazooka, and his shout 'I got me a tank!' was perhaps the most exultant sound to come from a general in any war. On one occasion he engaged a tank at point blank range with a pistol – that wasn't brave, it was just temper, he explained afterwards, but it expressed the spirit of the man.

If William Dean was not the most brilliant general the United States ever produced – he made many mistakes and admitted them – he was surely the bravest. An expedition in search of water for the wounded was rarely a chore carried out by a three-star general, but Dean was doing just that when he fell down a slope and was knocked unconscious. He spent thirty-six days wandering in the mountains and was eventually captured by the North Koreans – weighing four stone lighter than when he first arrived in Korea.

As the remnants of the 24th Division continued their retreat across the Kum River on July 13th, a cold wind was blowing round the flanks of the rear area officers and men in the port of Pusan. A sturdy and belligerent Texan clapped a hand like a leg of mutton on the shoulder of a young soldier surprised in the act of lighting a surreptitious cigarette when he should have been unloading stores. He eyed the young man sorrowfully for a moment and then seemed to sweep the whole of Southern Korea with a single all-embracing glance. Then he said: 'There's going to be some changes around here, son.'

Lieutenant-General Walton Walker (nicknamed Johnny after a proprietary brand of Scotch) had arrived in Korea as overall commander of 76,000 men – 18,000 American and 58,000 South Koreans.

Clearly, it *was* time for some changes. The armies in Korea could use a good man and they had got one.

In 1918 Walton Walker had commanded a machine-gun company in France; the 1930s found him in command of an infantry battalion in China; in the Second World War he was one of General George ('Blood and Guts') Patton's Corps commanders, and much of Patton's ferocious fighting spirit had rubbed off on him.

He was no desk general, and abhorred paperwork in any form – 'What the hell do I have a staff for?' was his explanation of this particular antipathy. His rare excursions into written directives were of the Winston Churchill 'Action this day' pattern.

He believed in showing himself to his soldiers on the ground as much as possible: he left the day-to-day administration of his headquarters to his staff officers.

Walker's impact on American and South Korean ground forces was immediate and astringent: unit commanders never knew when or where that questing, square face and bustling, intrusive personality would appear next.

The truth of the matter was that Walker was smarting under the undistinguished performance of some of the men under his command. He decided that the 1st Cavalry and 24th and 25th Infantry Divisions had done more than their fair share of retreating, and he told them so with uncompromising directness. But he brought good news too: a division of marines was on the way, bigger and better naval forces were coming, and bombing by B 26s and B 29s would soon be on the increase.

There would, said Walker, be no more 'bugging-out': there was no line behind them, and any further retreat meant that they would finish in the sea. Capture by the enemy would be the ultimate disgrace and worse than death: if some of them had to die, then they would die together. It was good, old-fashioned soldiers' talk, without gimmicks.

[5]

The men of the 3rd Battalion of the 23rd Infantry Regiment had had plenty of battle experience, and their appearance on September 5th, 1950 defied anyone to think otherwise.

They had gone to war as boys, but now they were, unquestionably, men: this showed in their faces, which had the greyish tinge of men who have lived with death for many days and have become accustomed to the sight and the smell of it. They had been shelled, machine-gunned and sniped; they had seen their comrades killed and they had killed men themselves.

In the battle for Taegu they had heard nothing but bullets and shellfire and mortars; the moans of wounded and the shouts of 'medic'; the shouts of their officers and NCOs.

Now, suddenly, they heard a new sound, which caused many of them to shake their heads in bewilderment. It was a high-pitched wailing noise; a sad sound, but it had a note of defiance about it – a sound which had been heard wherever the fighting was most savage for nearly two centuries: on the North-West Frontier of India, in the Western Desert, in France, Italy and Malaya.

GIs of Scots ancestry – a fair percentage in any American outfit – explained to their mongrel cousins that this noise signified that some limey outfit was coming to relieve them, that the tune the piper played was 'Hielan' Laddie', and that the relieving force, in the extravagant phraseology of the *Saturday Evening Post* of August 1950, was 'a kilt-wearing, proud, Scottish regular army organization'. The 1st Battalion The Argyll and Sutherland Highlanders had arrived in Korea . . .

PART TWO

The Woolworth Brigade

SEPTEMBER 1950–MARCH 1951

THREE

IN pup tents and foxholes, regimental command posts, divisional, corps and army headquarters and over vast tracts of North America the plaintive wail was heard: 'How long have we got to go this war alone?' It was a fair question: by the first week in August the United States Army in Korea had suffered something like 15,000 casualties, and the lists, under the coldly impersonal headings killed, wounded and missing were still coming in.

The United States seethed with resentment. Bereaved relatives, not a few of them belligerent and hysterical, stormed into local draft boards. And to Harry Truman, who had sent them to Korea, every man killed represented a deep, personal loss.

Forgetful of Great Britain and the Commonwealth, who had stood alone from 1914 until 1917 and from September 1939 until December 1941, the absence of British help in the Korean War represented the most deplorable deficiency of all.

Like a cocktail party which starts with a manageable number of guests, but which later becomes a riot with gatecrashers, nations jostled one another to get to the Korean War first.

Commonwealth naval forces had been at battle stations on the reef-strewn west coast of Korea from the beginning: the Royal Navy, represented by the light fleet carrier *Triumph*, the cruisers *Belfast* and *Jamaica*, the destroyers *Cossack*, *Consort*, *Comus* and *Cockade* and the frigates *Black Swan*, *Alacrity* and *Hart* were ranged alongside the United States Seventh Fleet. They were joined by the Royal Australian Navy destroyer *Bataan* and the frigate *Shoalhaven*; the Canadian destroyers *Cayuga*, *Athabaskan* and *Sioux*; and the New Zealand frigates *Tutira* and *Pukati*. Already 7,000 officers and men of the Commonwealth were at action stations.

The Royal Australian Air Force had been in the war from the start. When the North Koreans crossed the 38th Parallel on June

25th, 77 Mustang Fighter Squadron was in Japan as part of the Occupation Force and under the operational control of the 5th American Air Force. Officers and airmen had their wives and children on the well-appointed station at Iwakuni. Beer was 3d a glass and cigarettes twenty for a shilling. Domestic help was cheap and efficient, and household chores had become a thing of the past.

Flying duties were scarcely onerous: routine surveillance patrols and anti-smuggling watches, which amounted to little more than an aeronautical picnic. Homesickness is as prevalent among Australian servicemen as those of any other nationality, but there was little of this malady in 77 Squadron.

It was certainly not afflicting Sergeant-Pilot Raymond Treblico, who was on base operations duty on the morning of Sunday June 25th: Treblico was not thinking of Melbourne, but of the cold beer to which, at 12 noon, he would, by the clock, be entitled. The telephone had not rung for half an hour and it seemed unlikely that it would: amber frosted liquid pouring into a tall glass was as good a subject for concerted thought as anything.

Treblico's sybaritic reverie was interrupted by the harsh jangling of the telephone. Some joker checking up to make sure I haven't gone for that beer yet, he thought as he answered it.

'The North Korean Commies have crossed the 38th Parallel into South Korea,' announced an excited American voice without preamble. 'Boy, are things buzzing around here. I guess this means war.'

'What does?' said Sergeant Pilot Treblico in justifiable bewilderment – no one used the word war any more.

'Just this,' said the American, 'the 5th Air Force is alerted as of this minute, and that means you too. You got that?'

Treblico says that he had intended to say 'Roger', but in fact said 'Jesus Christ' instead.

By the afternoon of June 29th, 77 Squadron was fully operational. Although the hard-pressed American and South Korean ground forces did not know it at the time, they were receiving Australian help from the air.

Mastery of the sea and the air was all very well, but ships and aircraft did nothing to fill the gaps in the wavering line round

Pusan: it was wavering because, in spite of General Walker's forceful directives on the subject of retreating, the tendency of some American units to 'bug-out' had not been totally eradicated.

Carriers, destroyers, fighters and bombers were fine in their place, but here the desperate need was for guns, tanks, and foot soldiers – particularly the latter, for the war in Korea was to be as much an infantryman's battle as had been the Somme, and the conditions were not so very different.

Offers of ground forces poured in at an encouraging rate: Great Britain offered the immediate addition of an infantry brigade with the promise of more to follow; Australia, Canada, and New Zealand, ever generous with their fighting men in both World Wars, pledged infantry and artillery; the Turkish Government offered land forces – 'we can sure use some of those killer boys,' was the enthusiastic comment of a senior American staff officer on receipt of this piece of intelligence; France, Belgium, Greece, Ethiopia, the Philippines and Thailand promised contingents; South Africa offered a fighter squadron and India a field ambulance. Suddenly everyone wanted to get in on the act.

The Commonwealth Division did not yet exist, but the period of gestation was not to be a long one . . .

[2]

It can fairly be said that Great Britain's participation in the Korean War dated back to a June evening in 1949.

In the War Office Major-General Geoffrey Evans, Director of Military Training, threw the last of six files into his 'Out' tray and marvelled, not for the first time in his twenty-nine years' service, at the amount of paper which could be wasted at the dictates of so few; the files bristled with flags – 'Immediate', 'For Attention', 'Urgent', 'Confidential', and 'Top Secret'. They could all easily wait until the following morning, but Evans always liked to clear the 'In' tray by 5.30 PM; it always filled up fast enough the next day.

It was a Friday, and General Evans was thinking pleasurably of a round of golf, when his telephone rang. It was the Military

Assistant to the Chief of the Imperial Staff, and the CIGS wanted to see Evans immediately.

General Sir William Slim was wearing a grim smile when Evans came into his office; since this was habitual with Slim, Evans was not unduly perturbed. These two were old friends; in Burma Geoffrey Evans had been one of Slim's star divisional commanders.

It seemed that he still was. 'Hong Kong for you, Geoff,' said Slim. 'You're taking 40th Division out to bar the way to the wily Chinese. It's all right, you'll get some golf there . . .'

Since the end of the war, Hong Kong had slipped back to its placid peacetime way of life. The bathing beaches, the parties, the Repulse Bay Hotel, the Nathan Hotel and the seven-page menus in the Parisian Grill separately and collectively went to make up a near-military paradise.

The menace of Red China was not mentioned in polite circles, any more than the menace of Japan had been in 1941. But Red China was, in fact, making threatening faces in the direction of Hong Kong. 'Bill' Slim had seen enough of military unpreparedness in the heartbreaking defeat in Burma in 1942, and was making sure that Hong Kong was not caught napping again.

By the end of 1949 the New Territories had undergone a fantastic transformation. Two years earlier a single infantry battalion, the 1st Devons, stood alone between Red China and Kowloon; in 1948 the 2/6 Gurkhas and 1/10 Gurkhas took over when the Devons left for Singapore.

The Gurkhas departed for Malaya, where there was tangible evidence of Chinese Communist menace: rubber planters and tin miners were being murdered and a state of emergency had been declared. Experienced eyes looked to Hong Kong in 1949, and it seemed clear that if China wanted Hong Kong, then she only had to cross the border and take it.

In the view of General Evans they were welcome to try, once he had got his troops into position. There were now seven infantry battalions in the New Territories: the 1st Argyll and Sutherland Highlanders, the 1st Middlesex, the 1st Royal Leicesters, the 1st King's Own Scottish Borderers, the 1st King's Shropshire

Light Infantry, the 1st South Staffords. There was no bravado about Geoff Evans, because a soldier with three DSOs won in the Western Desert and Burma, had no need for self-advertisement. But when Evans said: 'Let 'em come,' he said so because he sincerely felt that the men of his 40th Division were equal to anything that Red China might throw against them, and there was not a man under his command who did not support him in this view.

Evans had no ground for complacency – he had been in too many tight spots in Burma to indulge in such a luxury – but he *did* have grounds for confidence: his men *were* ready for anything, and the efficient speed with which they manned defensive positions defied even the most searching military critic to think otherwise. It was Evan's boast that he could have every man under his command, fully armed and ready for the worst that Red China could throw against them, in two hours from the sounding of the first alarm. This was practised, often as many times as twice in a week, and at any time of day or night, by the old-fashioned method of a bugler blowing the alarm. Antiquated it may have been, but it worked.

Put a British battalion of infantry in a sybaritic peacetime station, where no war threatens and the most daunting military crisis is a defective refrigerator or a hold-up in mail from home, and it will get soft and unfit for war. In this dubious frame of mind British infantry took on the Japanese in Burma, Malaya and Hong Kong and emerged from this ordeal with varying degrees of distinction. In the main they fought valiantly: in khaki drill shorts, sweaty hosetops and nailed ammunition boots, East Surreys, Loyals, Manchesters, Gloucesters, Middlesex, Argylls, Royal Scots, Gordons, and Cameronians had only guts and bayonets to combat an enemy who had never heard of weekend passes or ammunition returns and whose training allowed him to regard the jungle as a friend rather than as an enemy. During this unhappy period British generalship savoured of 1916 and 1917, Britain's least productive years in leaders, and Japan acquired slave labour on an unprecedented scale.

But put that same battalion down on a bleak and inhospitable countryside; send it on long, toughening marches in the hot afternoon sun; set it to long and backbreaking hours of digging and wiring; chase it up and down mountains on the same soul-destroying exercise for weeks on end; feed it well and plentifully, brown it in the sun, ration its drinking to a few pints of ice cold and gaseous San Miguel (universally known as San Mig) beer per week, and restrict its carousing to one glorious blind until 23.59 hours on alternative Saturday nights; tell it at intervals what it may have to fight against, and get the chaplain to explain that Chinese Communism is something evil and directly opposed to the teaching of Christ – then, and only then, have you a taut and efficient fighting machine. Every battalion commander in 40th Division knew that this was expected of him, and if he ever forgot it Geoff Evans was constantly on hand to remind him of his responsibilities.

One brigade in General Evans' division was in a special category of preparedness: 27 Brigade, consisting of the 1st Argyll and Sutherland Highlanders, the 1st Royal Leicesters and the 1st Middlesex, and commanded by Brigadier Aubrey Coad, had been known since 1948 by the impressive title 'The United Kingdom Strategic Reserve'. In simpler language, this force was at instant readiness to go anywhere in the world at ten days' notice, and for this reason was known as the Fire Brigade.

Aubrey Coad accepted his hair-trigger role philosophically: twenty-two years in the British Army had convinced him that things are rarely what they seem, and he considered himself immune to any surprise that the Army might spring on him. He was (and is) a keen gardener, with an outdated but ingrained distrust of any vegetable unless it is grown by himself. During the months of waiting for action in Colchester he read the paper each morning, and having assessed the military situation, came to weighty decisions whether to put in cabbage or lettuce seeds.

The command of such a force demanded an exceptional fighting man, and the Military Secretary's Branch had made a wise choice in Aubrey Coad: he was no stranger to war, and had two DSOs to prove it. More important still, he had seen active service at all levels – he had been a senior subaltern in the pre-war

Palestine troubles and a captain at the outbreak of war; he had taken a battalion, and later a brigade, from the D-Day beaches to the upper reaches of the Rhine. His three battalion commanders – Lieutenant-Colonels Leslie Neilson of the Argylls, George Hutchins of the 1st Royal Leicesters, and Andrew Man of the 1st Middlesex had similar credentials, and if you had told them that in character, outlook, and background they were as indistinguishable as four hard-boiled eggs, they would not have felt unduly insulted.

And that went for the other battalion commanders of 40th Division – John Macdonald of the 1st King's Own Scottish Borderers and Vernon Barlow of the 1st King's Shropshire Light Infantry. They had all been through the upper-class forcing houses of public school and Sandhurst, and joined the Army in the military moribund period of the late 1920s. On joining their regiments they were eyed with sharp distaste by colonels and majors who had fought Boers, Dervishes, and Pathans at the turn of the century; they were reviled, victimized, castigated, humiliated, and unofficially court martialled by adjutants and senior subalterns who wore the Mons Star.

The Second World War, like the first, was a great social leveller. To their surprise, they acquired a liking and respect for company commanders, who had been commercial travellers and bank clerks in peacetime, and platoon commanders who had failed Certificate A (if they had ever heard of it), played soccer at weekends, smoked Woodbines, and called the drawing-room a lounge.

These colonels fought alongside volunteers, who had enlisted under age and over age or in a fit of patriotism; 'Belisha Boys' flung into the army before the outbreak of war; unwilling conscripts, who had never volunteered for anything and weren't going to start now. They got to know that new entity, the wartime battalion, and if they ever longed for the spacious days of 1930 they seldom said so.

But in 1948 they encountered a completely new military hybrid – the National Serviceman. Whatever their trade, background or potential value to Great Britain, young men of 18 were seized by a voracious British Army: patrician young men, just out of public school and headed for university and illustrious posts in industry, the arts, law, the Diplomatic Service, sheltered country life, and the peerage; middle-class young men destined for what were called 'white collar jobs' – junior civil servants, bank clerks, schoolteachers; ambitious young artisans – mechanics, metallurgists, sheet-metal workers, draughtsmen, dietitians, bricklayers, hairdressers; farm labourers, stable lads, apprentice jockeys, and embryo professional boxers and footballers; unskilled labourers, miners, barmen, waiters, spivs, drones, young men who lived on their wits, fairground hands, potential confidence tricksters and burglars, and self-confessed layabouts were relentlessly sucked into the British Army – willingly, thankfully, sulkily, mutinously, resignedly, tearfully, adventurously; with or without the approval of their parents and Members of Parliament . . .

For many of these boys – old enough to kill and be killed, but too young to vote – National Service proved to be a less daunting ordeal than they had expected. The British Army, like members of an exclusive club welcoming temporary members of dubious origin, received them with wary condescension. Initially they were not popular, because the regulars knew that they were in the Army but not of it: they had joined because the Government had told them to, not because they wanted to.

A small percentage of National Servicemen, frequently to their own surprise, found that they had been credited by Selecting Boards with qualities of leadership and were recommended for Officer Training Schools and National Service commissions. The system of weeding out was the subject of continuous criticism and constant revision, but on the whole the system worked.

The regular British soldier has always accepted his postings at home or overseas philosophically, and the National Serviceman soon learned to do the same: if the luck fell one way, they stayed in England and loudly condemned square-bashing, polishing, and white-washing of gate posts as a stultifying waste of time, which it was. Or perhaps they found themselves in Germany, where the routine was approximately similar, except that the liquor was more potent and plentiful, the cigarettes cheaper, and the women more accommodating.

If it fell the other way, they were called upon to swelter in

Singapore, Aden, the Canal Zone of Egypt, East and West Africa, and Khartoum. A fortunate few found themselves in the West Indies, where they observed millionaires at play and found that rum and fresh limes did not really count as an alcoholic drink.

A large number went to Malaya, where there were no millionaires, rum or fresh limes – only thick jungle, insects, snakes, and Communist guerillas: guerillas were rarely seen, but if they were they had to be killed with deadly thoroughness.

About two thousand of these went to Hong Kong and managed to make history before they came of age.

If you asked a battalion commander of the period what he thought of National Servicemen, you had to be prepared for a variety of answers: one would not instinctively approach Lieutenant-Colonel Andrew Man, Commanding Officer of the 1st Middlesex. He is a small man, but in stature only; in his quarter of a century of service, he found that there were many good days for small men, as his DSO, OBE, French Croix de Guerre, American Legion of Merit, and three mentions in dispatches testify.

Andrew Man has the conventional impassive senior officer's face behind an aggressively bristling moustache. As a battalion CO he was a disciplinarian of the older school, who did not suffer fools gladly.

Not greatly skilled with words, his pronouncement that 'it is up to the Commanding Officer to be relentless in maintaining the highest standards of leadership and man-management in his battalion for the benefit of the Army, and of the Regular and National Service soldier alike', smacks of the training manual. When he goes on to say that 'it is up to all of us soldiers, assisted by the Press, and by those who have the necessary authority, to go on showing the fathers and mothers of our youth that the Army system is all right, for indeed it is', the unknowing might be excused for muttering darkly about Colonel Blimp. Andrew Man would not care what you thought, anyhow . . .

Some scorn was poured on National Servicemen stationed in Great Britain. These youths could see little point in their enforced period in khaki, and many of them said so loudly and persistently – on occasions to Members of Parliament; the efforts of some of them were perfunctory at best. But no one, either in the early days in Hong Kong or since, has spoken disparagingly of National Servicemen in the presence of Lieutenant-Colonel Andrew Man.

Of the other infantry battalions in Hong Kong, the 1st Argyll and Sutherland Highlanders, with their fierce and touchy *esprit de corps*, were predominantly regular soldiers and their CO, Lieutenant-Colonel Leslie Neilson and the Adjutant, Captain John Slim, with an illustrious name to uphold, were happy that this should be so. But the 1st KOSB, the 1st Leicesters and the 1st KSLI all had something like 50 per cent National Servicemen in their ranks and felt no qualms thereby . . .

[3]

For the infantry battalions in the New Territories it had been a hard week, and one of those seemingly interminable exercises had taken up most of it: the six infantry battalions had run up and down the mountains and marched incalculable distances; they had practised the advance to contact, company attacks, and fighting patrols; they had dug endless trenches and filled them in again. Shoulders were sore from the cutting straps of web equipment, rifles seemed to weigh a ton and boots were uncomfortably full of feet.

The South Staffords had theoretically been wiped out by an ambush by the Argylls and the KOSB had suffered a similar fate at the hands of the Middlesex.

Majors wearing the white armbands of umpires strode about platoons and informed bewildered subalterns that due to their carelessness they were dead; the platoon commanders listened respectfully, secure in the knowledge that they would get it in the neck from their company commanders who would in turn be demolished by commanding officers.

The old soldiers swore horribly and told the young soldiers that the Western Desert and Burma had been cushy compared to this. The young soldiers secretly hoped that they too might be declared dead by an umpire, but hypothetical death was a dubious distinction because General Geoff Evans had seen through that one a long time ago, and usually 'corpses' were put on to digging.

On the evening of Friday August 18th, 1950, Brigadier Aubrey Coad sat at slippered ease in his Kowloon flat. In spite of some dreadful tactical errors which he had observed at first hand on the exercises – he felt at peace with the world: the whisky and soda at his elbow was long, strong, and cold; he had a first-class brigade, although he rarely let them think so; an encouraging odour from the direction of the kitchen told him that 'Cookie' was doing his solemn best to produce a palatable dinner. Best of all, his wife had recently joined him from England.

He knew, however, that there was little enough cause for casual complacency: the news from Korea was bad and getting worse almost hourly. The United States expected help from the British Army and expected it quickly.

In emphasis of this, the telephone rang and General Evans told Coad that he would pick him up at 8 AM next morning and take him to Headquarters, Land Forces. Coad permitted himself a small, interior sigh: it looked as if the 'Fire Brigade' was in business at last.

That 27 Brigade *was* in business was made abundantly clear to Coad by the Commander-in-Chief, General Sir Robert Mansergh. Whereas it had formerly been at ten days' notice, it was now denied this luxury: Coad's orders were simple and categorical – he was to command a force consisting of the 1st Argyll and Sutherland Highlanders and the 1st Middlesex and take it to Korea on the following Friday. Were there any questions?

There were several hundred, but Coad realized grimly that this was neither the time nor the place to ask them. He saluted, and sought out his battalion commanders. It was next door to a physical impossibility to get two infantry battalions ready in all respects for war in five days, rendered doubly so because the British Army, running true to form, had elected to prepare for war over a weekend. Consequently the bars of Kowloon teemed with soldiers of both the Argylls and the Middlesex; in many a club and private house the officers were being entertained. Sunday stretched before them in sybaritic idleness; they did not know that Aubrey Coad, after several hours of intensive work with his Brigade Major, Douglas Reith, was asking some trenchant questions of Headquarters, Land Forces and getting what seemed to him at the time some remarkably silly answers.

He was to take a brigade to Korea at seven – no, it was only six now – days' notice?

'Exactly.'

But why only two battalions, and weak ones at that? Why not the whole brigade – Argylls, Middlesex, and Leicesters?

'The Leicesters can't be spared – if they went we'd be too thin on the ground in the New Territories.'

Wasn't this a job for predominantly *regular* soldiers? The Argylls were almost all regulars as were a good percentage of the Leicesters. On the other hand, the Middlesex were over 50 per cent National Servicemen.

'The orders say the 1st Argylls and the 1st Middlesex.'

Coad sighed and tried again. It wasn't much of a force to take to a full-scale war.

'It's all we can spare. Anything else?'

Artillery?

'You're not taking any – they say the Yank gunners are pretty good.'

Transport?

'You're not taking any of that either. Won't need any – the Yanks have got a vehicle to about every five men.'

Rations?

'You'll be all right there – turkey, for every meal.'

'Sounds to me like a Woolworth Brigade.'

'That's just about what it is. Damned good luck to you, Aubrey . . .'

By 6 PM on Saturday August 19th only Coad and his two senior staff officers in 27 Brigade knew what was happening. The two battalion commanders were to be told on Saturday; company commanders were to be kept in ignorance until Sunday.

Aubrey Coad had some scathing comments to make on what he considered to be excessive security measures. 'I am as security minded as the next man,' he declares, 'but in this case I think secrecy was carried to extremes. If a field officer commanding a company cannot be relied upon to keep a secret, then he has no right to hold the rank he does. Comparatively junior staff seemed to know about the whole show from A to Z, but the integrity of the chap who has actually got to do the job seems to be suspect.'

Lieutenant-Colonel Leslie Neilson was told as he got ready for a dinner party; he returned to the Argylls, still in his dinner jacket, with an injunction to his jeep driver to drive like hell without having an accident. On arrival back to his lines he sought out his second-in-command, Major Kenny Muir, and his Adjutant, Captain John Slim, and they went into a huddle which lasted far into the night.

That Saturday the 1st Middlesex were *en fête*. The unit recreational transport, known as 'passion wagons' had been filled to overflowing with as many men as could be spared for the occasion – a swimming gala against the Hong Kong Volunteer Corps, the men of a dozen nationalities or more who made up the Colony's territorials.

The Volunteers had wiped the floor with the Middlesex in the relay races and diving; the Middlesex had their revenge at water polo.

The scene round the swimming baths was traditional and everlasting and a million miles removed from war: the officers in their tight blue overalls with the thin red stripe down the side and 'bum-freezing' white mess jackets with miniature decorations; their ladies in ball gowns, for there was to be a dance afterwards.

The rank and file, with no mountains to climb or trenches to dig the following morning, roared encouragement and cheerful abuse at their comrades as they strove to catch the torpedo-like swimmers of the Volunteers, and compared their diving to the performance of seals after buns. But they let out an ear-shattering roar of approval when Corporal Charlie Small clinched the water polo match with a shot which would have decapitated the Volunteers' goalkeeper if he had not ducked in time.

The prizes were presented; the inevitable soldier, under the frosty eye of Regimental Sergeant-Major Sam Weller, fell into the bath fully clothed. Then Andrew Man, amid a burst of applause stepped forward to make a closing speech.

He did not get very far with it: outside the swimming pool a worried-looking staff officer from Headquarters, Land Forces, told Andrew Man that the 1st Battalion the Middlesex Regiment (Duke of Cambridge's Own) were at five days' notice to embark for Korea on the following Friday.

Major Aymar Clayton of the 1st Middlesex had stayed Saturday night with a friend, Major 'Puggy' Pugh, a staff officer at Headquarters. At 11.30 AM on Sunday morning Pugh was pouring remedial brandy and ginger ale. Pugh knew all about the projected move to Korea, and hoped only to get Aymar Clayton on the road back to health before the news hit him. He was only partially successful: midway through the second restorative 'horse's neck' the telephone rang. The telephone call told Aymar Clayton that playtime was over and it was time to go to war . . .

Back in England, Major John Sloane of the 1st Argylls had every reason to feel pleased with himself. He had just completed a course at the Joint Services Staff College at Latimer – a congenial course of instruction which taught soldiers, sailors and airmen to work together in harmony. On its conclusion, a successful student was entitled to place the small letters *jssc* after his name in the Army List (sometimes irreverently said to stand for 'just slightly senior to Christ'). The course had been professionally valuable, interesting and, mainly thanks to the sailors and airmen also attending, convivial.

There was still better to come. Sloane was returning to Hong Kong with his wife and two young sons, prep school men both, with an insatiable urge to travel.

But this was not to be. A curtly worded telegram told Major Sloane that he would return to Hong Kong forthwith – and alone . . .

Lieutenant Sandy Boswell, of the 1st Argylls, was also on a

course, but of a rather more exacting type: in the sweltering heat of Singapore, he was mastering new and terrible ways of killing. At the end of the course he was looking forward to a fortnight's leave with his parents in the comparative cool of Kuala Lumpur. A cable marked 'urgent' very soon put paid to that.

All through that long hot Sunday the officers of Headquarters, 27 Brigade, the 1st Argylls and the 1st Middlesex prepared for war – without flurry and without fuss; soberly, quietly, methodically, as men who have made it their trade.

Soldiering was their trade, and a sudden move such as this had been practised over and over again: there had been alarms and excursions during the 'Fire Brigade's' period of readiness in England, and they were used to them. All that they had lacked to date was an active hostile enemy; soon they would meet one.

The younger captains and subalterns went about their duties with scarcely concealed excitement; the senior officers, who had been to war before, were not men with a surfeit of imagination. But nearly all of them admitted to a slightly cold feeling in the region of the stomach, inevitable in men who, only a short five years before, had heard the unnerving crump of shells, the spiteful whine of snipers' bullets, the cries of men terribly wounded . . .

The 'Midds' and the 'Jocks' were not officially in the know on Sunday, but even the most unobservant private realized that something was brewing. There had, of course, been the usual rumours: they were going to Malaya, they were going home (this one was severely discounted by almost everyone), they were going to march on Red China.

But they were obviously going *somewhere*, and before very long: you didn't find officers working on Sundays, certainly not this late – they played tennis, swam, larked about with the girls on the beaches or got boozed up in the Officers' Club, the Hong Kong Hotel, the Parisian Grill or the Repulse Bay.

The Officers' Mess waiters of both battalions – Officers' Mess waiters are always watchful men, who miss very little – observed that both Adjutants, Captains John Slim of the Argylls and Nick Carter of the Middlesex, bolted down their dinners as if it was their last meal on earth and fled hotfoot to their offices immediately after. There was no sign of either Commanding Officer; the company commanders and subalterns turned up much later than usual for pre-dinner drinks, and there was no demand for the popular six o'clock 'snifter' at all.

Most sinister of all the orderly-room clerks, who everyone well knew were an idle lot of pen-pushers who knew a cushy billet when they saw one and seemed to belong to some sort of trade union, were actually *working* after dark.

[4]

By 9 AM on Monday August 21st, 1950, every man in the New Territories knew that the 1st Argylls and the 1st Middlesex were going to Korea. The 'bush telegraph' worked in a variety of ways, but it was only a bare three hours ahead of the BBC in London. At midday an indecently emotional voice said 'British troops are going to Korea', and speculation promptly ended.

Both the Argylls and the Middlesex were sadly under strength. In order to make them up to approximately six hundred each, men would have to be found from the KSLI, the KOSB, the Royal Leicesters and the South Staffords. This news was particularly well received by Lieutenant-Colonel Hutchins, Commanding Officer of the Leicesters, who had been asking Brigadier Coad at approximately half-hour intervals why his battalion was not going. Now, at least, some of them certainly would be.

Later in the Korean War, the *Evening Standard*, a number of indignant mothers, and several Members of Parliament became excessively heated when it was learned that 19-year-olds were fighting in Korea.

Lieutenant-Colonel Andrew Man (who rarely read the *Evening Standard*) had quite another problem: it was telling the 18-year-old National Servicemen of the 1st Middlesex that he could not take them, for a War Office order had decreed that no soldier under 19 would be sent to Korea. As a result of this, the Middlesex needed another 150 men. The 18-year-olds of the Middlesex considered this to be ridiculous and said so vociferously. More – they said so to the Commanding Officer in person.

There has always been a certain amount of rigmarole for a private soldier who seeks a private interview with his Commanding Officer: the first application goes to his Company Sergeant-Major, who examines the validity of the applicant's case. Having surmounted this daunting hurdle, the company commander asks the most searching questions. Then, some three days later (if he is lucky) the enterprising soldier finds himself on Commanding Officer's Orders, assuming that he has had his hair cut. The RSM eyes him from head to toe in steely silence for fully two minutes; then, to the accompaniment of a screeching cadence of sound, he delivers the sweating and by now tongue-tied soldier to the Presence.

On this Monday morning the 18-year-olds of the 1st Middlesex slashed their way through red tape and thumbed their noses at King's Regulations. They spotted the CO on his way to the orderly room, and closed in on him in a solid phalanx; it was as well that RSM Sam Weller was busy elsewhere. Man remembers one young soldier's 'I'm pretty sure I *am* 19, sir,' as a gem of its kind.

Andrew Man, disciplinarian and fierce upholder of the uttermost limits of military protocol, stared fixedly at the ground and found that he had some difficulty in speaking.

'I'm sorry,' he said, when the clamour died down for a second, 'but those are the orders.' Then he strode to the orderly room, upbraided three men quite unnecessarily for sloppy salutes, and cursed himself for a sentimental idiot.

In all, some 250 men had to be found from the other four battalions. Some Commanding Officers would have gleefully off-loaded the 'dead wood': the idle, the dirty, the insubordinate, the frankly impossible and what RSMs call ''orrible things'. But Aubrey Coad would have none of this: he stipulated firmly that the men 'milked' from the Leicesters, KSLI, KOSB, and, South Staffords, should be exclusively volunteers. He need not really have bothered . . .

Volunteers . . . since time immemorial the old soldiers have counselled the young never to volunteer for anything; in the Second World War they displayed the natural perversity of the British soldier by volunteering in droves for the Parachute Regiment, the Special Air Service, the Commandos and the Special Boat Service.

But in the early days of a recruit's service, volunteering could have dire consequences.

The 1st Battalion the King's Own Scottish Borderers stood on parade in the blistering sun. On the previous Friday they had been tired, dusty and sweaty and anything but smart at the end of the exercise. Now every man wore freshly laundered shirts, shorts with knife-edged creases, blancoed belts, highly polished boots and a glengarry.

The KOSB did not like the idea of this particular parade and half an hour earlier had been saying so with mournful truculence: the Midds and the Argylls were going to Korea, and here they were on a bloody drill parade.

Their Commanding Officer, Lieutenant-Colonel John MacDonald, twenty-three years a soldier, six feet tall and conventionally moustached, eyed them with approval. At first sight, when he took over command in Edinburgh in 1949, he had not been over impressed: MacDonald had come from the more astringent atmosphere of West Africa, and it had been very apparent to him that the 1st KOSB had been performing guard and ceremonial duties in Edinburgh for too long – in this respect they were not unlike the first American troops to go to Korea from Tokyo.

But eighteen months in Hong Kong had done wonders for them.

RSM Lees had a parade-ground voice which was surely audible in Red China. He took a deep breath and bawled: 'PAY ATTENTION!'

Here it comes, thought the Jocks morosely; first the wee mon'll gie us a bollicking, then he'll march us into the bloody ground.

'VOLUNTEERS FOR KOREA! ONE PACE FORRRWARD MARRRCH!'

With one accord, every single man on parade took a pace forward. The RSM turned about with a crash of iron-clad heels, faced the CO and saluted wordlessly.

The CO smiled a small, secret smile. He had privately considered the parade to be a waste of time, but half an hour's drill wouldn't hurt them . . .

[5]

Men of the Argyll and Sutherland Highlanders and the Middlesex had been the first to land in France in 1914; they had been in the van again in 1939. It was appropriate enough that they should, in the words of Frank Owen of the *Daily Express* – a man always ready with a telling phrase in time of war – comprise the Third British Expeditionary Force.

Thirty-six years had passed, but the mood was the same. The mood in the 1st Middlesex was one of satisfied jubilation, except among the 18-year-olds who felt themselves to be the victims of some gigantic confidence trick.

The mood of the Middlesex was the mood of No 4 Platoon of 'B' Company, and by the same token the mood of Second-Lieutenant Christopher Lawrence, just 20 years old and six months out of Sandhurst.

Together with Second-Lieutenants Pat Woolocombe and Stanley Fothergill, he had joined the battalion three months earlier. They had been very young, very new, very apprehensive, very anxious to please, very smart: their peaked caps were rather stiff in the peak and sat dead straight on their heads; their new tropical uniforms looked as though they had been hanging in the window of Gieves of Bond Street only the week before. Like Andrew Man, twenty-three years earlier, they were eyed with sharp distaste by Adjutant and Commanding Officer and dispatched to their platoons. They were received with cheerful resignation by hard-case platoon sergeants, who had seen them come, go and die in the last ten years or so, and wary condescension by hard-swearing young men who came from places like Hayes, Barnet, Edgware, Southall, Uxbridge and Ruislip and parts of London which rarely figure in the itinerary of an American tourist: Stepney, Shoreditch, Bow, Hackney, Lewisham and Bethnal Green.

Christopher Lawrence was the archetype of all keen young

infantry officers – the colonel's, and private's dream subaltern. His whole life evolved round No 4 Platoon: he boxed with it, ran cross country with it and played football with it; he congratulated it, admonished it and listened to its complex stories of amatory and alcoholic endeavour. He not only knew every man in it; he knew every man's size in boots, leadership potentialities, religious denomination and domestic problems.

He soon learned that Sergeant Frank Butters would have been a Company Sergeant-Major years before, were it not for an inordinate thirst for beer – this suited Lawrence well enough, for gallons of 'San Mig' made not an atom of difference to Butters' speech, gait or immaculate bearing on early morning parade. If Butters' road to further promotion was barred, then so much the better for Christopher Lawrence and No 4 Platoon.

He discovered that Corporal Joe Pentony had the biggest repertoire of what are sometimes known as 'smoking-room stories' in the whole battalion, and that Corporal Charlie Small, a muscular young man with a weakness for amateur theatricals, had been 'Mr Junior London, 1948' and had flexed his muscles on the stage for the fearful enjoyment of thousands of young women.

There were Privates Sam Buckle and Chalky White, who needed to be watched carefully nearly all the time, but particularly at weekends: they had almost, but not quite, perfected a plan for smuggling themselves aboard the married men's transport – the married state carried with it the privilege of not having to be back in camp until first parade on Monday morning; Buckle and White considered this to be unjust and outdated.

There were Johnny Ridge and 'Greg' Peck, first-class soldiers both, but chronic grousers. Christopher Lawrence knew them all as an old married man knows his wife.

Lawrence liked No 4 Platoon and No 4 Platoon liked their commander. It was a state of affairs which augured well for battle, and battle was not so very far off . . .

Private George Farmer of Bow, London, E, was typical of the permanent Londoner – the virile, cunning and cheerful Cockney who has filled the ranks of the Middlesex Regiment for over a century. The George Farmers of this world frequently swear that they would rather starve than do a steady job; they speak in a staccato voice and prefix every noun with the same tuneful and copulative adjective; they are unscrupulously dishonest in their dealings with the rich, recklessly generous to anyone poorer than themselves.

They speak the strange and barely comprehensible tongue of the East End: if they have no money (as is frequent) they are hearts of oak (broke) or boracic lint (skint). A suit is a whistle, the wife is the trouble, the stairs are the apples, a Jew is a front-wheel skid, feet are plates, and so on, up and down the English language.

George Farmer had a problem. He had only another month left to serve, and a calendar above his bed with only thirty days still to cross out, testified to the imminence of his return to England – twelve months away from Bow was enough for any bloke.

When he heard the news about Korea, George Farmer betook himself to the Company Sergeant-Major. He considered himself to be a patriot and a good soldier, but enough was as good as a feast. 'I'm due for home, Sar-Major,' he said.

'You *were* due for home,' corrected the CSM unsympathetically. 'Any questions?'

The mood of the 1st Argylls was predominantly belligerent – a frame of mind which had stood them in good stead as the Thin Red Line at Balaclava, at Le Cateau, and in Malaya in 1941, where the 2nd Battalion had marched across the Johore Causeway into Singapore ninety strong, with a solitary piper playing 'A Hundred Pipers' at their head.

Successive generations of young men in Great Britain are accused by their elders of a lack of the spirit of adventure. No such accusation could be levelled at the men of 40th Infantry Division; and whatever his motives, every single officer and man in the New Territories wanted to go to Korea.

No one man wanted to go more than Company Sergeant-Major 'Wullie' Robertson of the 1st Argylls. Man and boy, he had served with the regiment for twenty-four years – on the North-West Frontier of India, North Africa, Italy, North-West Europe and Palestine. But now it seemed he was too old to go to war again; middle-age, that insulting period in a soldier's life, had caught up with him.

Robertson whipped himself into a white heat of indignation and even threatened to stow away on the ship when the Argylls sailed. He had a friend and ally in the Regimental Sergeant-Major, Paddy Boyde – a comparative youngster with a mere seventeen years' service behind him and a Distinguished Conduct Medal won at the crossing of the Rhine. But the RSM had troubles of his own: some 'thievin' Chinee' had stolen his trews . . .

It was the proud boast of Sergeant 'Butcher' Bailey, medical sergeant of the 1st Middlesex, that he could detect a malingerer at five hundred paces: he knew all about the men who developed coughs and colds just before night exercises, the opportunists who limped into the sick bunk just before route marches and the real experts whose all-consuming ambition was to perform before a specialist.

But now he was encountering malingerers in reverse: men who were 'excused boots' (but nevertheless performed prodigies of valour on the football field) announced that their bunions, athlete's foot and ingrowing toenails had miraculously recovered; no coughs or sneezes were to be heard; diarrhoea and constipation had become things of the past. Private Ted Ellis, who had fractured his arm in a motor-cycle accident, stormed into the sick bunk, demanding that the plaster be removed from the limb immediately, if not sooner.

It is on record that two privates of the Argylls pondered long and seriously on the advisability of going out in the Kowloon bound 'passion wagon' on their last night of freedom. For word had got round that any men brought up on a charge laid by the Corps of Military Police would not go to Korea.

Both these Jocks had conduct sheets of impressive dimensions; both had crossed swords with the Military Police, taxi drivers who had demanded fares which they considered to be excessive and on one memorable occasion with a naval shore patrol (as they were both wearing sailors' caps and kilts at the time, naval intervention was perhaps excusable).

That night these two Jocks eschewed the bright lights of Kowloon, with its many pitfalls for the unwary and the concupiscent, and so they went to Korea . . .

As the Argylls, Middlesex and Brigade Headquarters worked feverishly to get ready for war, an old troopship, like an elderly but recently painted streetwalker on a familiar beat, was pursuing a leisurely course from Singapore to Hong Kong.

There was rarely anything particularly colourful about a troopship: they were severely functional vessels, their decks thronged with soldiers leaning over the side, smoking, sun-bathing, sleeping, playing cards and housey-housey.

But the top decks of this particular trooper were colourful, because the female passengers had all been on a spending spree; blending with the bronzed torsos and drab khaki shorts of the soldiers were gay summer dresses in every conceivable striped, spotted and floral pattern. The ship's ladies' hairdresser had done a roaring trade. It was clear that, so far as outward appearances were concerned, the ladies were leaving nothing to chance. Some had acquired a fine tan; others, new to the tropics, had lingered too long in the sun's rays and were painfully pink and peeling. But none of them cared, for every turn of the propellers was taking them nearer to their menfolk.

There were children, too, who raced about the decks, got lost, were sick in the dining saloon, cannoned into stewards carrying trays of drinks and generally made a nuisance of themselves. But mothers and crew members forgave them all, for this was a Great Adventure and Daddy, that funny man who came to see them for brief periods, would be on the dockside to meet them and the family would be all together again.

Many of them were wives and children of men of the Argylls and the Middlesex.

Regimental Sergeant-Majors are rarely credited with having

hearts, or even human feelings; they are almost godlike beings, who must be obeyed instantly and avoided whenever possible.

RSM Sam Weller of the 1st Middlesex had a heart, although this was not often apparent to the men of the battalion, and it was heavy within him. No one in the battalion knew this, for the only thing that Weller wore on his sleeve was the insignia of a Warrant Officer Class One.

When he got married in 1936, he had been a sergeant: young for his rank, keen, efficient, ambitious. During his courtship he had made sure what his fiancée was signing on for: he explained to her many times that the British Army was as hard a taskmaster to the women as it was to the men, and often more so; that whether the husband be general or private, heartbreak, frustration, resentment against the military system, separation and disruption of settled family life are always just around the next corner.

And so it had been. In 1939 Weller had gone to France with the 2nd Battalion. On his return from Dunkirk, he had snatched a brief weekend with his wife before going to South Wales. This was followed by a spell in the North of Scotland which lasted well into 1944. On June 6th of that year the 2nd Middlesex were among the first assault troops to land in Normandy and, with Sergeant Weller inevitably in the van, fought their way to the upper reaches of the Rhine.

Then it was Palestine and Egypt, both at the time non-family stations. In 1948 Mrs Weller packed two trunks, four suitcases, disregarded the tearful protests of her two small daughters when they learned that they could not take the cat, secured four pigtails with brilliant red ribbon and survived a cancelled sailing to Germany.

In 1949 the Weller family were reunited at Mill Hill Barracks, and Mrs Weller became halfway to being convinced that she was a married women. Then the overseas posting hammer fell again, and this time it was Hong Kong.

In 1950 families joined soldiers serving overseas on a 'points' basis; that is to say, points were awarded for length of service overseas and period of separation. Sam Weller had more points than anyone else in the Middlesex Regiment, if not in the entire

British Army. When family passages were allocated to the 1st Middlesex in July 1950, the Weller family were in a commanding lead.

On Wednesday August 24th Weller was doing the hundred and one things that fall into the RSM's province before a move – checking men, weapons, stores and ammunition. On that same day Mrs and the Misses Weller, aged 7 and 13 and with their pigtails still secured by red ribbon, stood at the rail of the troopship and chanted in unison: 'Only four more days to Daddy.'

The British Army had beaten the Weller family yet again – by a clear seventy-two hours . . .

There was the story of the two corporals of the South Staffords.

It is not a story of heroic endeavour, although it well might have been if they had had their way. It is the simple story of two men who just wanted to fight.

They had both been in action before, and one of them wore the Military Medal. Their combined service totalled something like twenty years, and they had been with the Staffords, one of the battalions in Mad Mike Calvert's brigade, in the desperate and bloody offensive on Mogaung – one of the worst and most terrible battles of the Second World War.

They were neither of them large, but there was about both of them the tough and stocky durability of the men from the Potteries. In common with almost every other NCO and private in 40th Division, they had volunteered for Korea. Their names had not been selected, and they were fed up – like CSM Robertson of the Argylls, they were in the mood to stow away on a ship – any ship – which was going to Korea.

Sergeant Bert Cleland of the 1st Middlesex sat in the sergeants' mess, morosely contemplating a bottle of San Miguel Beer and wishing it was a stout and mild. He was wondering whether to have another or go to bed, when a Chinese mess servant came in. 'Two men outside,' he announced.

'What d'you mean, two men outside? Are they sergeants?'

'Not sergeants.' The waiter held up two fingers.

'Corporals?'

'That's right,' said the waiter. 'Two stripes.'

The elected spokesman of the two – the one with the Military Medal – told Cleland that they most desperately wanted to go to Korea, and were prepared to go as privates if necessary. Was there any chance?

'I don't know what they thought I could do for them,' said Sergeant Cleland, 'but I couldn't send two blokes like that away, even though I did know that the battalion was up to strength.'

So Cleland sought out the orderly officer, who was sympathetic but scarcely more forthcoming. There was nothing that he personally could do about it, except take their names and promise to bring them to the notice of the CO when the question of reinforcements arose. 'Someone's going to stop a bullet soon,' he prophesied grimly. 'You might find yourselves there quicker than you expect.' And with that the two corporals had to be content . . .

Now the volunteers from the KSLI, the Leicesters, the KOSB and the South Staffords were absorbed into their foster-parent battalions: KSLI, Leicesters and South Staffords were accorded a riotous welcome by the Middlesex. The Argylls greeted their lowland cousins, the King's Own Scottish Borderers, who were some species of Scot who did not wear the kilt, with wary condescension.

The Third British Expeditionary Force was ready.

[6]

There had been nothing like the departure of the 1st Argylls and the 1st Middlesex from Holt's Wharf, Kowloon, for half a century.

The British soldiers of the 1890s were dispatched to subdue the enemies of the Queen Empress, clad in khaki drill uniforms and wearing helmets designed for almost any purpose except protection from the murderous heat. They were hastened on their way by weeping relatives and cheering crowds waving patriotic Union Jacks.

If the spectators – and they comprised every man, woman and child who could get there – did not wave Union Jacks, they did everything else. For this was an old-fashioned embarkation, with all the trappings of peacetime trooping.

There had certainly been nothing like it in either world war, when security-veiled transports crept away like thieves in the night. But in Hong Kong on this day security had been joyously abandoned; the move to Korea had been successively classified Top Secret, Secret, Confidential and Restricted. Now, just one week later, it had been de-restricted with such gay abandon that the whole world knew which units were going, where they were going, when they were going, and in which ships.

There were staff officers in starched khaki drill, civilians taking time out from their offices, women in gay summer frocks and carrying parasols. There were farewell scenes, as husbands and wives kissed desperately and tiny, pudgy fingers explored sun-burned faces and playfully tweaked noses; but there were few scenes like this, for the 'Third British Expeditionary Force' was predominantly bachelor, although some of the families were now just forty-eight hours away from Hong Kong and heart-breaking disappointment.

The drabbest people on the quayside were the men the excitement was all about; awaiting the order to go aboard the ships, they stood in patient lines of jungle green, reasoning not why. They had been soaked with rain as they waited to board the trains in the New Territories. Now that the rain had stopped, they were soaked with sweat as they waited on the wharf, weighed down by ammunition pouches, haversacks, water bottles, big packs, kitbags and rifles. But not a single man looked dejected, as they drank appreciatively from the bottles of iced orange squash provided by a local soft-drinks manufacturer.

In addition to being cheered, and photographed and interviewed by the numerous reporters, the Argylls and Middlesex had been addressed by General Sir John Harding, Commander-in-Chief Far East Land Forces.

John Harding employed no gimmicks; a short, dapper figure, in conventional khaki drill and red tabs, he talked to them as he thought generals should talk to soldiers: 'Shoot quickly, shoot straight and shoot to kill,' he told them uncompromisingly; 'don't go away unless you are ordered to do so; stay and fight it out. Even if the enemy gets behind you or round your flank, stay put. So long as you have ammunition, water and a little food – stand and fight.'

It was good, tough soldiers' talk, and understood by every man present.

And if any soldier was in doubt about the purpose of British intervention in the Korean War, Malcolm MacDonald, High Commissioner for South-East Asia, was there to remind them of their responsibilities. He left them in no doubt as to who was the real enemy.

Among all the uniforms, MacDonald looked a little insignificant – hatless and dressed in a slightly crumpled tussore suit. But his voice and mannerisms, strongly reminiscent of his father, impressed even the solidest and most inattentive private.

'You will be fighting North Koreans,' he told them, 'but the weapons they are using are Russian. Their training is Russian. This is the same enemy who threatens France and Britain.

'This is an historic expedition. You will be fighting as if on the soil of France or on the beaches of Britain. The Korean War is part of a Russian attempt to conquer the world and make slaves of us all. This aggression is being resisted not only by individual nations, but by the free peoples of the world under one organization, of which you are a part. It is up to you to show the world the valour and unconquerable spirit of the British people.'

They were sombre words, but the pipe band of the 1st King's Own Scottish Borderers and the brass band of the 1st Leicesters were there to disperse any undue solemnity.

The drones began as the bladders of the bagpipes filled with air. Amid rapturous cheering, the pipe band of the KOSB, in full dress, marched, and counter-marched to *Hielan' Laddie*, *A Hundred Pipers*, *The Cock o' the North*, and *The Campbells are Coming*. The band of the Leicesters listened to this primitive music with Sassenach tolerance, waiting for the pipers to run out of breath, Then, not to be outdone, they launched into Sousa marches, selections from Gilbert and Sullivan and *Oklahoma* and, with a nice sense of timing, *Slow Boat to China* and *This is the Army*, *Mr Jones*.

The Argylls filed up the gangways on to the blue-grey cruiser

Ceylon; the Middlesex boarded the pea-green aircraft-carrier *Unicorn*. Having stowed their kit, the soldiers thronged the decks. They had assumed the dress which the British soldier in the tropics likes best – stripped to the waist and with caps comfortably on the back of their heads. They whistled and sang to the music and proffered advice of a ribald and spurious nature to the faultlessly laundered military policemen on the quayside, who smiled bleakly in response.

Boatswains' pipes shrilled; Royal Marine buglers called ships' companies to attention. The Leicesters' band struck up *Auld Lang Syne*; the pipers, inevitably, capped this with *Will ye no' come back again?* Gangways were closed and disconnected; slowly and reluctantly, the crowds began to disperse. Pipers and bandsmen, breathing hard and sweating profusely, refreshed themselves with cold drinks.

Then the *Unicorn* and the *Ceylon* slipped down the harbour towards the Lei Mun gap and the open sea . . .

FOUR

FOR Brigadier Aubrey Coad the early days in Korea had represented a period of awed and bewildered fascination, during which time he strove to familiarize himself with United States Army procedure.

Coad had flown into Korea ahead of his brigade, accompanied by his Brigade Major, Douglas Reith, Colonels Man and Neilson and a skeleton staff. From the moment he touched down in a shabby Dakota on a dusty, run-down airstrip to be greeted with a refreshing lack of rank consciousness by a young air force captain who said, 'Jesus, are we glad to see you guys,' he had lived in a world which, to the conventional British military mind, was one of fantastic make-believe.

It was a world in which transportation, cold tea, Coco-Cola, dog-faces, crap games, stripes worn upside down, bug-outs, foul-ups, real estate (a favourite expression of senior American officers for hill features), egg-beaters (American for helicopter), chow lines, goofing-off, chicken colonels, combat zones and rain checks were supposed collectively and separately to make sense.

He had an early meeting with General Walton Walker, and after the first crushing handshake they inspected one another warily. Coad saw that there were four stars on Walker's helmet and on his shirt collar, and a quick calculation told him that he only rated one. Walker saw a tall, well-built soldier with prematurely grey hair, a prominent nose and a determined chin. He personally had a liking and respect for 'Limeys', which had been engendered in France in the Second World War. And even if this guy Coad didn't wear as many medal ribbons as some veteran privates first class, he had clearly seen some combat and obviously knew the score.

He was whisked from one divisional headquarters to another, and discovered what a regimental combat team was; he marvelled at the number of staff officers: in a British Division there are rarely more than six lieutenant-colonels on the headquarters staff, but in an American division there may be that number of 'bird' colonels. Coad soon lost count of the number of majors and captains.

Coad had considered that the security arrangements in Hong Kong before the move to Korea had been excessive. But the Americans in Korea seemed to go to the other extreme. There must, thought Coad wryly as he was photographed for the twentieth time, be a happy medium somewhere.

The war in Korea was probably the least understood and most reported war in history. There were reporters and photographers everywhere: they raced in jeeps to the line companies, with pockets stuffed with candy bars for Korean children; they swarmed into briefing conferences at divisional headquarters and regimental command posts; they appeared in chow lines, intent on personal interviews with the guys who had been slugging it out up front; they did human interest stories on motor-pool sergeants who had adopted orphan Korean children and chaplain's assistants who had built altars out of K-ration boxes.

There were the tough round-the-world 'by-line' correspondents in the Ernie Pyle tradition, who had sweated out beach landings from Anzio to Tarawa; there were keen young men, too young to have covered the Second World War, who made it their life's work to ensure breakfast-table shell-shock from coast to coast. Tense-faced radio network broadcasters came back from a hundred different little 'hells' and went on the air with fresh, hot-off-the-bunker stories of heroism for a hundred North American radio stations. There were agency men and re-write men, who shared billets with staff officers and re-shaped pieces hastily hammered out in regimental command posts under fire. There were those rarefied beings, the 'name columnists', writing 'think' pieces about what war does to men and how war brutalizes some men and makes others more gentle. There were correspondents who had never heard a shot fired in anger and never intended to, picking their copy from a big board in the briefing room, and other correspondents who made up their stories from a fertile imagination under faked date lines. There were women correspondents, wearing GI pants and steel helmets, trying to look feminine and warlike at the same time.

On the night of August 28th, Coad went to bed, his brain reeling under the impact of half-formed plans, orders, counter-orders, optimistic reappraisals of the conduct of the war and prognostications of disaster.

But the next day would be different, and he would be able to get to work on a soldier's job. On the 29th his brigade was due at Pusan.

[2]

Many places in the universe, where filth and squalor predominate, have competed for the dubious title the 'fundamental orifice of the world'. Basra, Port Said, and Calcutta are all strong contenders – Basra, indeed, has even been described as being 80 miles up it – but in the opinion of the more widely travelled veterans of the Korean War, the port of Pusan carries off any palm.

But on August 29th, 1950, Pusan was looking its best, although its characteristic excremental smell was still depressingly apparent. For this day was an historic one: it signalled the arrival of the first British troops to fight in Korea – the 1st Argyll and

Sutherland Highlanders, and the 1st Middlesex, once known as the Fire Brigade and now dubbed the 'Something for God's sake Brigade'.

For the soldiers it had not been a pleasure cruise, and each day afloat had been devoted to training: both ships vibrated to the rattle of firearms, firing into the water from the stern; officers lectured the men on the more obviously disgusting aspects of Communism; there was strenuous physical training. Fortified by the best food they had eaten in twelve months or more, secure in the rightness of their cause, bronzed by the sun, hardened by the rigorous training in the New Territories, there was not a man in either battalion who would not have taken on ten North Koreans single handed or signed on for a twelve-year stretch.

There were, of course, the usual crop of shipboard rumours – the same sort of rumours that had swept the decks of troopships in the Second World War: the war was almost over; they were only going to Korea as a propaganda force to show the Union Jack; Pusan was even now under shellfire. A vintage three-badge able seaman aboard *Unicorn*, who had been at Dunkirk and had a low opinion of all things and persons military, prophesied an early evacuation from Korea – 'always getting the b— pongoes out of trouble, that's us'.

The Australian destroyers, *Warramunga* and *Bataan*, which had escorted the convoy, signalled 'Good luck, Pommies', and swept away like greyhounds released from the traps, intent on more warlike pursuits.

Then the *Unicorn* and *Ceylon*, with crew closed up at action stations and with the two regimental flags fluttering alongside the White Ensign, drew smoothly into the sunlit quayside.

If the send-off from Hong Kong had been flamboyant, the reception at Pusan was positively frenetic.

It was a glittering pageant of international goodwill. On the quayside was John Muccio, the United States Ambassador to Korea, flanked by senior American army officers; Sidney Faithful, the acting British Minister; senior South Korean officers, resplendent in gold braid; Brigadier Aubrey Coad, embarrassedly clutching a bouquet of gladioli which had been presented to him by the Director of Information, Doctor Helen Kim; representatives of the South Korean Government and members of the United Nations staff.

There were flags everywhere: the Union Jack, the Stars and Stripes, the Dove of the United Nations, and the sphere and bars of the South Korean republic. Children conscientiously waved miniature Union Jacks.

GIs stood on the quayside, swaying gently from heels to toes in the classic stances of cowboys in a Wild West saloon; their cameras were at the alert, their hands thrust deeply into their pockets. Every one of them chewed rhythmically on gum as they looked upwards in critical appraisal of the newcomers.

And looking at the Tam-o'-shanters of the Argylls: 'Say, what in hell sort of a hat is that?' (They hadn't seen the kilted pipers yet, and that really threw them.)

There were two bands; mercifully they did not attempt to play together. As the first Argylls came down the gangways, a South Korean navy brass band struggled manfully with a Sousa march.

The other band seemed to have stepped straight out of *Porgy and Bess*. They were all Negroes, and they stood at ease fingering their instruments with grinning expectancy as they waited for the last laboured bars from the Koreans. At their head stood a gigantic drum-major who, from the top of his gleaming helmet to his shining size 13 boots, must have measured seven feet.

There was a Korean girls' choir in navy-blue skirts and spotless white blouses – surely the cleanest things that Pusan had seen before or since. Each one held a piece of paper on which was written the words of *God Save the King*; they had learned the tune from a hymn in their schoolbook entitled *God is our refuge*. The choirmaster, an earnest and profusely sweating little man wearing spectacles, had got the words from Brigadier Coad who, to his eternal shame, could not remember all the words of the second verse. Their throats suitably moistened with American chewing gum, the girls rendered the National Anthem in shrilly piping chorus.

Now it was the turn of the Negro band. They played *Tiger Rag* and *St Louis Blues* and then, with a baffling reversal of mood, *The Chicago Tribune March* and *To the Shores of Tripoli*':

this was a delicately turned compliment – if the United States Marines could not be there, then the British 27 Brigade was the next best thing.

These selections were played standing to attention, but the *pièce de résistance* was at hand. The drum-major had clearly kept up the pose of austere dignity for too long, and it was slipping fast.

To the Shores of Tripoli ended with a crash of drums and cymbals and a blare of trombones. Then the drum-major's self-control deserted him entirely: his ebony face split into a huge grin and he let out a yell of sheer enjoyment. This was his big moment.

To the strains of *Colonel Bogey*, played in exaggerated rag-time, the Negro band broke into a march approximately twice as fast as the British rifleman's quickstep. They were headed by the drum-major, twirling his baton aloft and practically galloping with his knees all but hitting his chin. The soldiers cheered and whistled; children screamed with joy, frantically waving their Union Jacks; Brigadier Coad fiddled self-consciously with his bouquet – to him it seemed that one day of fantastic make-believe had quickly given way to another.

But the show was not over yet, and by popular acclaim it was stolen by Pipe-Major McGlynn and Pipe-Corporal Pitkeathly, both wearing kilts, who climbed on to the after-gun turret of HMS *Ceylon* and introduced the bagpipes to astonished Korean ears. A shapely American nurse goggled in disbelief at the kilts, turned to her neighbour and said: 'Now, isn't that the cutest thing you ever saw?'

Nor were the press idle. Photographers jostled one another for position and two of them, memorably, came to blows. A wag of the Middlesex offered his sten gun to the smallest one, who had just been floored, camera and all, by a shrewdly delivered right hook. 'Reckon you'll need this, mate,' he said sympathetically.

The reporters, however, found it was not their day. Before their departure from Hong Kong, General Harding had told the soldiers to keep their ears open and their mouths shut, and nobody was willing to give anything away . . .

The party was over, and now it was time to go to war. The trains, as a grim reminder that war was an uncomfortable and sordid thing, stood at the sidings. Between the men of the Argylls and the Middlesex and the sailors of *Ceylon* and *Unicorn* there was a most heartening *rapport*: the sailors took the view that they had brought the pongoes this far and might as well take them a little further – they cheerfully carried kitbags, ammunition boxes and stores with the soldiers. The transfer of men and equipment from ship to train was, in the words of Brigadier Coad, 'a truly combined operation'.

The sailors entered into the spirit of the thing with such *élan* that it would not have surprised the Argylls or the Middlesex to find at least one sailor in every platoon.

No single officer or man of the Argylls or Middlesex wishes to remember that train journey: some of the 'old hands' had experienced austerity class rail travel in India, and rated it Pullman class compared to this one. Seating accommodation was hard and cramped; there was no air, only choking dust; sleep was impossible, except in positions of appalling intimacy. The carriages moved in a succession of bone-shaking jerks. The trains laboured up even the most gradual of slopes and frequently stopped for no attributable reason.

Major Jim Gillies, commanding 'C' Company of the 1st Argylls, left his compartment to investigate one of these mortifying delays. He discovered that the train had stopped for the simplest of all reasons: the engine driver had become tired and was fast asleep by the side of the railway. Gillies, among whose many virtues patience had never been the most prominent, stirred him with his boot accompanied by threats uttered in a hideous mixture of Hindustani and Arabic. The driver misunderstood the instructions but not the method of persuasion, and the train creaked painfully on.

The rest of the journey to the front was faster, but scarcely more comfortable. Lorries, driven by Negroes who smoked, chewed and sang melodious spirituals, took them at breakneck speed over lamentable roads to a harbour area near Kyongsang.

And there they made camp and awaited the call to action.

[3]

In the camp near Kyongsang, 27 Brigade found that the Korean War so far represented no great hardship. They were, admittedly, still the 'Cinderella' Brigade – they had no transport and no artillery, only rifles, machine guns, mortars, bayonets, and guts – but they found that Prince Charming, in the shape of the United States Army, was determined to keep the Wicked Squire and the Ugly Sisters at bay until their turn for action came.

The Quartermaster Branch showered rations upon them with a carefree abandon which would have earned any British supply officer an immediate court martial: there were grapefruit, pork slices, and canned tomatoes for breakfast; tins of hashed meat appeared in a bewildering variety – the main ingredients seemed to be chicken, hamburger meat, turkey, veal, and spaghetti; there was fruit juice by the gallon and instant coffee by the ton. Non-smokers in the British Army are in a minority, but even allowing for the one man in ten who was not an addict, there were seldom less than thirty a day per man – Lucky Strikes, Chesterfields, and Camels. There was no bread, but quantities of what the United States Army calls 'hard bread' – biscuit or 'hard tack' to the British soldier.

In 27 Brigade the men – particularly the Argylls – liked to start the day with porridge, thick and hot, with sugar (Middlesex) and salt (Argylls); they got acres of cereal instead. There were fruit bars, oddly reminiscent of liquorice allsorts, and toilet paper of the soft texture only to be found in the more luxurious North American hotels. An Argyll private, asked by an officer if he had any complaints about American rations, said: 'Aye, surr, too much f— turrrkey.'

American weapons experts called on both battalions and demonstrated the bazooka with driving salesmanship. 'This is a real honey,' declared a sergeant, a fanatical enthusiast, to the men of the Middlesex. Private Greg ('Grouser') Peck discovered that he was an unerring marksman with this weapon and Second-Lieutenant Christopher Lawrence appointed him bazooka man for No 4 Platoon. This caused Peck's grousing to reach a new crescendo: the bazooka, plus the bombs, weighed approximately twice as much as the load carried by any other man.

In spite of the almost leave-camp atmosphere in 27 Brigade – the distant grumble of artillery was the only reminder that there was a war on – the overall military situation was causing grave anxiety to General Walton Walker.

Substantial reinforcements had arrived: to the battered 24th, 25th, and 1st Cavalry Divisions had been added the 2nd Division, which had come direct from the United States, the 5th Regimental Combat Team from Hawaii and a regiment of Marines. The Americans had deployed in Korea the equivalent of five divisions, but they were all on an improvised basis and had, as a matter of grim necessity, been committed to battle piecemeal. Some semblance of order had been restored to the South Korean divisions and they were scarcely recognizable as the bewildered and demoralized rabble which only a short month before had been in headlong retreat.

The bridgehead round the port of Pusan was perilously small in area, but its perimeter was far too long for the troops available to hold it. From Pusan to Masan, the western extremity of the line, was only 45 miles. In all, 150 miles of line had to be held by a force which was woefully depleted in numbers.

Taegu, the last major town in South Korea, had fallen and American and South Korean troops were disposed on a defensive arc, based on the Naktong River, covering the approach to Pusan. On the left, in the coastal area to the south, was an American Marine Regiment; then the four American Divisions in the order 24th (in which was incorporated the 5th Regimental Combat team), 25th, 2nd, and 1st Cavalry. The South Korean Divisions were on the right, or eastern flank.

On the face of it, the portents seemed reasonably good: with complete command of the sea, the United States could pour in men and supplies virtually unhindered; the USAAF and the Royal Australian Air Force were strafing rolling stock, bridges, truck convoys, and other lines of communication south of the Yalu River boundary with Manchuria, thus taking the sting out of the Communist advance and slowing it down considerably.

The heavy weight of aviation thrown against the North Koreans was equal to any thrown against any army anywhere in the Second World War. Having no air force of their own, the Communists had clearly never planned a war in which air action would play a decisive part.

But although these unremitting air attacks slowed down the enemy advance, they did not stop it. When air bombardment drove the North Koreans off the good coastal roads they took to the sheltered inland tracks. In this sort of country aircraft, tanks, and heavy transports were at a disadvantage, but the North Koreans solved the problem by using horsecarts, handcarts, pack animals, and human carriers – particularly the latter: a twelve-year-old boy could handle a belt of cartridges; a woman could, and frequently did, carry a dozen grenades. By day Communist troops hid in the railroad tunnels which honeycomb the Korean hills; by night they plodded steadily on, unobserved by the road-bound Americans, who always travelled by road and never paused to think that any army would do otherwise. Consequently, American units not only had the enemy in front of them; they were frequently behind them and on their flanks as well.

Anxiety eased towards the end of August as fresh American troops came in – a most welcome reinforcement appeared in the form of four battalions of medium tanks. But in spite of this build-up in ground strength and persistent attacks from the air, the unpalatable fact remained that the greater part of South Korea was in enemy hands.

General Walker, anxious as he was to plug the sagging line with the newly arrived British troops, had agreed with Brigadier Coad that 27 Brigade should not be committed to action until their transport arrived. But on September 1st the North Koreans launched an offensive all along the line, a last all-out attempt to breach the Pusan perimeter. There was no question of waiting for transport now: every American, Englishman, and South Korean who could hold a rifle was needed.

A lesser man than Aubrey Coad might have boggled at being responsible for 18,000 yards of front with two weak battalions.

Coad's front was on the Naktong River, south-west of Taegu. There was a gap of 4,000 yards between his left flank, which later became 9,000 yards. The greater part of a North Korean division infiltrated into this gap with insolent ease.

Intensive round-the-clock patrolling was the order of the day for the Argylls and the Middlesex in their early days in the line.

There were reconnaissance patrols – half a dozen men with orders to draw enemy fire and report back with information about enemy numbers and dispositions without getting involved in a scrap; fighting patrols of twelve or more – they were looking for trouble and usually found it: their orders were to meet the enemy and hit them hard with everything they'd got; standing patrols more commonly known as ambush parties – they lay in wait for the enemy and exterminated him at close range; patrols whose bounden duty it was to take *one* prisoner (but they could kill as many as they liked). It was on jobs like these that second-lieutenants really earned their meagre pay.

There was a patrol of the Middlesex on September 8th, a hum-drum enough affair as patrols go; it merited a single line in the battalion war diary and three inches in the British press which described it as a 'skirmish'. It was the usual thing: the hard, sweating slog into mountainous country and the brisk flurry of hastily aimed and inconclusive shots. This particular patrol was only notable in that one member of it was the first British soldier to be killed in action in Korea.

No one knew better than Colonel Andrew Man that you cannot make war without casualties: he had seen many dead men in Normandy and was to see many more in Korea. But the laconic entry in the battalion intelligence report, '1 OR killed', left him with a feeling of bereavement and deep personal loss which lasted a full week; after that, he admitted, a certain merciful callousness set in.

Private Reginald Streeter was in no way exceptional, either as civilian or soldier: he was a National Serviceman, aged 19½; he had been a plumber's mate and came from Guildford; he had a girlfriend aged 17, who coaxed waves into the hair of Guildford ladies; they were not engaged, but hoped to be so when he returned home, for Reg had only three months left to serve in the Army; he listed his hobbies as boxing, football, and the cinema.

It seemed unlikely that Reginald Streeter would make history, but a stray bullet on a Korean hillside ensured that he did.

Just two days before this patrol Reginald Streeter, an indifferent correspondent, wrote to his mother. In one paragraph he wrote: 'We have got to show them who is boss out here. Some of us will be killed, one way or another. Don't worry about me, I shall be all right. I shall come back.'

Just after the postman delivered this letter in Worplesdon Road, Guildford, the traditional whistling telegraph boy, unwitting harbinger of death in every war, came with the telegram which began, 'The War Office regrets to inform you . . .'

On the same day a patrol of the Argylls ran into trouble. This, too, was a reconnaissance patrol, commanded by Captain Neil Buchanan. Stories of atrocities perpetrated on American prisoners by North Koreans had filtered down to 27 Brigade: men had been found tied up and shot through the back of the head, bayoneted, and mutilated. Whatever the outcome of the patrol, Buchanan impressed upon his men, any wounded were to be brought back.

The patrol duly drew the fire they expected and split into two parties. Sergeant Alexander Walker and three privates were slightly wounded during the first exchange of shots. Then the volume of enemy fire increased, disclosing the strength and disposition of the enemy.

Buchanan saw that he was outnumbered by something like three to one. His orders from Colonel Neilson had been clear-cut: pin-point the position of the enemy, estimate their strength, and get out of it. He had four wounded, and the North Koreans were closing in for the kill. 'Our job's done here,' he told Sergeant Walker. 'It's time to go.'

At that moment a burst of machine-gun fire hit Neil Buchanan above and below the waist and in the right leg. He shared this burst with his batman, Private 'Tam' Taylor, who took a bullet through the shoulder.

It was clear to Sergeant Walker and every other man of the

patrol that Buchanan was mortally wounded. Even if they did get him back, with the most expert medical attention immediately available he could not be expected to live longer than an hour.

Buchanan ordered the patrol to go back, but Sergeant Walker realized that only an act of physical violence would get Taylor away. Between officer and batman, Walker knew, existed a David and Jonathan-like relationship: they had started the war together and meant to finish it together.

Taylor settled his officer as comfortably as possible, took a long swig from his water bottle and after a prolonged search found a crumpled cigarette . . .

[4]

Coad's 27 Brigade were getting the feel of total war, and the Argylls and Middlesex slid into their new role as effortlessly as well-drilled extras in a Hollywood epic. The two battalions were not, as one American correspondent enthusiastically described them, composed of hardened veterans of the Second World War who had been blooded in combat – apart from officers of the rank of captain and above, warrant officers and sergeants, none of them had ever been in action before. But their behaviour in the field kept the impression alive in American eyes and it never really died.

They had lived for a fortnight in sporadic contact with an enemy who overlooked them, shelled them, and mortared them – 'Dig if you want to stay alive' was an executive command heard in every platoon. There were more patrols, of the same rough nervous character as the ones in which Captain Neil Buchanan and Privates Taylor and Streeter had been killed.

Neil Buchanan had been popular with all ranks, and the death of such an officer always moves any battalion to a cold, fighting fury; this has always applied particularly forcibly to Scottish regiments. When the officer concerned has been done to death in circumstances of the most cowardly barbarism – Buchanan's and Private Taylor's bodies were later found, and it was clear that they had been butchered in cold blood – the mood becomes one which only bloodletting on a substantial scale will assuage.

This was the mood of No 1 Platoon of Major David Wilson's 'A' Company on the night of September 16th. It fell to the men of this platoon to pay back the North Koreans at an encouraging rate of interest.

At 3 AM on the previous morning, at an hour when man's spirit is at its lowest, a strong enemy patrol had attacked a machine-gun section, and although it was driven off it was not before it had inflicted casualties. On the 16th it was decided that this machine-gun section should move to another position inside 'A' Company's defended locality. It seemed that the North Koreans were determined to liquidate this section, and Sergeant John Morrison took out a patrol which duly reported an even stronger enemy party on the way. Showing a strange lack of originality, they came head-on at the position where the machine-gun section had been and found it empty.

The enemy were inside 'A' Company's position and would have to get out somehow before daylight – by Wilson's reckoning, this would be in the next half-hour. Wilson spoke to each platoon commander on the radio in turn, giving explicit orders that the enemy were to be most thoroughly exterminated.

'A' Company sat tight and waited and listened. Cocking handles of bren guns were drawn back, and change levers moved to 'automatic'; the trigger fingers of riflemen itched almost unbearably; bayonets were fixed, for this looked like a party which might well end with cold steel – an Argyll speciality.

Meanwhile, the North Koreans, balked of the coveted machine-gun section, blundered about in the darkness and chattered excitedly among themselves.

It was soon uncomfortably apparent that the enemy were all round No 1 Platoon's position, but daylight, with characteristic Korean perversity, refused to break through. Sergeant John Robertson of No 1 Platoon first uttered bloodcurdling threats to any man who fired before getting the order and then looked at the luminous dial on his watch – another five minutes at the outside . . .

Suddenly a North Korean hurled a grenade at the platoon position; it was followed by a second and a third. Not a sound came from Robertson's men, and now at last the murky dawn was breaking; dim figures, tightly bunched, targets to set a soldier dreaming, could be seen all round the position. 'Come on, you bastard,' muttered Robertson to the recalcitrant dawn.

As dawn broke Sergeant Robertson, never a soft-speaking man, shouted: 'Let 'em have it!' at the full pitch of his lungs. The men of No 1 Platoon needed no second bidding, for they had been waiting for this moment for an anxious, crawling hour; bren gunners ignoring an instruction in *The Small Arms Manual* which states that bursts should be fired three shots at a time, kept their fingers firmly on the trigger until the magazine was emptied; sten guns pumped bullets into the North Koreans, which the dawn had revealed in attitudes of the most foolish unpreparedness; riflemen fired as fast as they could work their bolts: it all represented a tremendous amount of metal in one place at one time, and it proved far too much for this target.

It was all over very quickly, and although it only merited two lines in the battalion war diary, it was a very considerable victory: the enemy did not fire one shot in reply – they ran in undignified panic from a battle which earlier had promised a very different outcome. Fifteen dead were counted on the position, and at least half a dozen other 'runners' were seen to fall and lie still on their way down the reverse slope. There were no Argyll casualties.

Back at company headquarters, when the noise had died down, Major Wilson inquired what had happened. 'Enemy patrol, sir,' said Sergeant Robertson economically; 'they were not liking it, so they left.'

[5]

Back in Tokyo, while the 8th Army licked its wounds and prepared for the next battles, General Douglas MacArthur was launching a massive operation which he declared would end the Korean War with a single bold stroke. D-Day was to be September 15th for the seaborne invasion of Inchon.

It was the sort of manoeuvre which had brought MacArthur such spectacular successes in the Pacific in the Second World War, and in its giant conception it was his biggest project to date.

The Navy, the Army, and the Marines raised a positive cataract of objections: the main approach to Inchon, called the Flying Fish Channel, was narrow and treacherous and seemed to invite the laying of enemy mines; dozens of small islands crowded the waterways and piled up the tides; between the channel and the shore lay a two-mile-wide sea of mud, which would not support infantry on foot. It would be a blood-bath; another Gallipoli or Tarawa. The odds against its success, declared the gloomiest critics, were 5,000 to one – and that was being generous.

General J. Lawton Collins, Army Chief of Staff, and Admiral Forrest P. Sherman, Chief of Naval Operations, did not like the idea and said so. Even MacArthur himself admitted that no operation he had planned in the past had been more fraught with danger. But it soon became clear to dissenters and supporters alike at the final debate in late August that MacArthur was going ahead with the amphibious invasion of Inchon and that they might just as well get used to the idea.

MacArthur stabbed at a map with the stem of his famous corn cob pipe. 'Inchon,' he said, 'is the anvil. Give me the anvil and General Walker's 8th Army will be the hammer which will smash the North Korean Army. The Inchon invasion will save 100,000 lives.'

FIVE

By the end of the first week in September the North Korean advance on the Pusan perimeter had shot its bolt. During their final and frenzied offensive along the Naktong River, the work of General Walker resembled that of a man trying to repair a dam which is undermined by continuous flood: he had to plaster up weak points in the line – and they were many – with every man who could fire a rifle (and even with some who couldn't), and maintain an ammunition and food supply at the same time.

During this fighting, which was always of the most savage and desperate character, the term 'bug-out', synonymous with panic and disorderly retreat, was all but buried for ever. Now it was replaced by 'we're gonna hold'. They held, and many of them died.

Now at last Walton Walker's 'stand or die' order was being obeyed to the letter: this perimeter war had suddenly become a war of many perimeters, with units cut off by the enemy's deep penetrations and left to fight it out alone: there were units, shattered by mortar and machine-gun fire without food or water, holding off wave after wave of North Koreans; there were units, ordered by commanding officers to hold to the last man and the last round, which did exactly that.

The United States 35th Infantry Regiment particularly distinguished itself, and won a Unit Citation for its gallant conduct. In one particularly bloody action, Sergeant William Jecelin was awarded a posthumous Congressional Medal of Honour: when ammunition was running out, he left the perimeter under heavy fire to gather weapons and grenades from enemy dead. He continued to fight even when paralysed from the waist down and, rather than jeopardize their safety, refused to be moved when the order to withdraw came. 'Just leave me a carbine,' he instructed the men of his platoon, 'and I'll take some of them with me.'

It was men like Jecelin and Private Melvin Brown, who conducted a one-man last stand with rifle, grenades, entrenching tool, and finally bare fists, who saved the Pusan perimeter.

But the prime architect of victory was Walton Walker himself. He called the North Korean offensive the last gasp of an over-extended enemy, but it was rather more than that: if it had succeeded it would have broken the 8th Army apart and brought the United Nations Command to a point of grave crisis.

Walker drove everyone hard, but no one harder than himself. 'Hell, doesn't the old man ever sleep?' muttered disgruntled regimental commanders, as the familiar belligerent figure with pistol at hip and shotgun in hand alighted from his armoured jeep and issued orders which seemed to range from the suicidal to the frankly impossible. Like many other successful generals, he probably made more enemies than friends, but the end justified the means.

On September 15th, Walker declared the Pusan perimeter

safe. The valleys of the Naktong River were littered with the bodies of North Koreans. When the fighting was over in the area behind the 35th Infantry, the regimental commander declared that the slaughter was on an even greater scale than at the Falaise Gap, where the German 7th Army had been destroyed in 1944.

Depriving his soldiers of some of the thunder, General Walton Walker said over and over again that it was air superiority alone that kept the precarious American foothold in Korea during July, August, and the early days of September: without it the defenders of the approaches to Pusan would have been swept into the sea. When Walker had mentioned Dunkirk to Brigadier Coad, he had not been indulging in melodramatics.

Mustang fighters and napalm were the weapons which enabled the hard-pressed Americans to hold the Pusan perimeter. When the war began, the United Nations air force had consisted of three wings of American jet fighters and the Mustangs of 77 Squadron, Royal Australian Air Force. The jets were found to be useless for the kind of air war that developed: built for 'dog fights' three or four miles above the earth, they could not carry rockets and napalm to bring their fighting power down to ground level. Consequently, the American jets were modified for close-support work and the American units changed over to Mustangs.

Napalm, the lethal mixture of aluminium soap, acids, and gasoline, added considerably to the beastliness of the Korean War. Together with rockets, the long, yellow, banana-shaped napalm containers roared down on the hillsides. Within seconds of impact they spewed out 1,000 degrees centigrade of liquid flame which left frightful destruction in its wake: buildings were razed to the ground; tanks became ovens and trucks blackened skeletons: men were grilled like barbecue steaks left on a turning spit for too long.

Anyone launching into a controversial discussion about Commonwealth participation in the Korean War must have his facts right: he may talk lyrically of the Argylls, the Middlesex, and the Gloucesters; the Royal Canadian Navy; the New Zealand gunners. But in the 'Surrey' in Fleet Street, where two or more Aussies are frequently gathered together, he must remember (if he wants to emerge unscathed from fierce argument) that 77 Mustang Squadron of the Royal Australian Air Force were the first in action and were still in there slugging at the finish.

They were a tough crowd, these Australian pilots – they had to be: it was not the toughness of the Battle of Britain men, who saved England but still brought an immoderate undergraduate spirit to the operation, nor were they tough in the fatalistic 'Here's to the next men to die' tradition of the bomber crews. Rather they were tough in the melodramatic manner of their American counterparts: they had a refreshing contempt for any sort of dress regulation, and set off on missions to the accompaniment of outrageous comment; on their return (if they returned) they indulged in a riot of song and story of such startling crudity that it even impressed the American flyers.

Harmony between Americans and Australians has never been one of the most heartwarming features of the English-speaking world: in the Second World War, in the absence of a tangible enemy, they frequently fought one another and relentlessly seduced one another's women. But in Korea, as mutual respect grew, jokes about big-mouthed Yanks and convict hulks from England began to lose their sting.

Between American and Australian pilots in the Korean War there grew a wonderful *rapport*: time and time again the reckless gallantry of Australian Mustang pilots, as they dived into lashing flak to pin-point targets, won the admiration of American flyers. 'Those guys certainly set the pace,' was the comment of Lieutenant-General George E. Stratmeyer, Commanding General of the United States Far East Air Force, and a man ever sparing with his praise for any non-American effort.

Launched into the war right from the start, 77 Squadron had an impressive contribution to record: during August, 40 pilots had shared over 800 sorties and flown 1,745 hours in combat operations; from the beginning of the war to the end of August, RAAF Mustangs had, with rockets and napalm, destroyed 35 North Korean tanks, 226 vehicles of varying sizes, 4 trains and a quantity of ammunition and fuel dumps.

Casualties were inevitably heavy: Squadron Leader Graham Trout was one of the very first United Nations airmen to die in Korea – he dived on to his target, the Samchok railway station on the East coast, and failed to pull out; Wing Commander Louis Spence, Commanding Officer of 77 Squadron decorated with the American Legion of Merit, went into a napalm attack from 700 feet in heavy cloud – he, too, could not pull out of his dive and his Mustang exploded on impact; Sergeant-Pilot William Harrop ran out of fuel and crash-landed on the banks of the Naktong River – he survived the crash, but was killed by North Korean guerillas. Fourteen pilots were to receive the American Air Medal for 'meritorious service in combat'.

New of Australian success in the air filtered through to the men of the 3rd Battalion, the Royal Australian Regiment in Japan. They heard the news with mixed feelings: these flying jokers were getting all the fun, and at this rate the war would be over before they ever got to Korea. Little did they know . . .

[2]

Even the most rabid critics of the Inchon invasion plan had to admit that the advantages of Inchon were manifold and obvious: it was the second-best harbour in Korea, and was only twenty-five miles from Seoul, with all the importance the city held, both psychologically and strategically. Near by was Kimpo the best-surfaced airport in Korea. All the main Japanese-built roads and railways crossed at Seoul.

MacArthur heard out every dissenter, displaying rather more patience than usual. But they were fighting a losing battle and they knew it. Sometimes politely, sometimes peremptorily, sometimes just plain rudely, MacArthur overrode all objections.

Clearly, it was a job for the United States Marines: MacArthur had made this plain from the outset, and he was in the habit of getting what he wanted. On this, at least, everyone was agreed: if the Marines could not pull it off, then no other fighting man of any nationality could.

In 1950 it was said in some cynical and unbelieving circles that the United States Marines had an unwarrantedly good opinion of themselves. They certainly believed, as they always had done, with a passion that almost amounted to religion, that there were no men on earth like them.

The Marines considered that the American Army was useless. But the Marines – they were wonderful: they were the toughest fighting men on God's earth, and just let any sonofabitch try to argue the point.

It was no bad frame of mind for a fighting man to be in, and not inaccurate: the proof of the pudding is in the eating, and the United States Marine Corps had always swallowed it whole: the Marines would not have been exceptional if every man in the Corps had not believed them to be the best.

The facts are inescapable. America's manpower in time of war was graded something like this: the navy and air force took a large share of the technicians – the men who joined 'Seabees', the army engineers, and air force engineers, were skilled men; they got top pay rates and usually served in comparative safety.

But of the millions of 'unskilled' youngsters, the Marine Corps had first choice. The Corps recruited boys who wanted to fight, the combative types, who were afraid that the war might end before they could destroy an enemy of the United States. Typical of these were the young men of the 1st and 5th Marines, earmarked for the assault on Inchon: they were mostly youngsters who had not had combat experience, but there was a fairly generous leavening of officers and NCOs of the type who had raised the Stars and Stripes on Iwojima, and were determined to do the same thing at Inchon. In fact, they did just that.

Although they swarmed ashore with their customary *élan*, the Marine assault on Inchon was scarcely in the murderous tradition of assault landings in the Second World War. The first objective, the island of Wolmi-do, had been so thoroughly pounded from air and sea that a Marine pilot described it as 'one worthless piece of real estate. It had been hit so hard and so long with so many things that it looked like it was quivering. I expected it to roll over and sink any minute.' Wolmi-do was secured at a cost of 17 wounded Marines; enemy casualties totalled 108 dead and 136 prisoners. Within one and a half hours of the first Marine setting foot on shore, the Stars and Stripes fluttered from the island's highest peak. The Marines were running true to form, and MacArthur sent out a message to the fleet: 'The Navy and Marines have never shone more brightly than this morning.'

The invasion of Inchon went off smoothly and according to plan. The first wave of Marines ran into rifle and mortar fire, but casualties were miraculously light – eight killed and twenty-eight wounded. Once the city was secure, three spearheads fanned out simultaneously: one eastward to Seoul, twenty miles away; one south towards Taejon, to link up with Walker's 8th Army; a third north-west to secure Kimpo airfield and Kaesong, on the 38th Parallel.

The North Koreans did not give up Seoul without a struggle, and they fought desperate rear-guard actions at every point of American entry. The Marines, who had thought that the invasion of Inchon was a cake-walk and hardly worthy of United States Marines' intervention, found that they had to fight from house to house, from block to block, from barricade to barricade.

On September 29th, General MacArthur announced that Seoul was in American hands.

[3]

As the Marines went ashore at Inchon, General Walton Walker was formulating plans for the break-out from the Pusan perimeter.

All portents seemed at first sight good for an all-out counter-offensive: the Americans outnumbered the North Koreans two to one and outgunned them six to one; the enemy had barely 30 per cent of their original forces left, and numbers had been made up by captured South Koreans unwillingly pressed into service – their reluctance to fight for their former enemies was such that there were frequent instances of men being forced into the line at gun point.

The Communists had lost half their equipment and food supplies, even by their frugal standards, were dangerously low. So too was morale, and it sank to a level of near despair when the news from Inchon, in spite of the efforts of the High Command to suppress it, started to filter through. The victorious days of July and September, when it seemed that all they had to do was throw the Americans into the sea, seemed very distant: now they were trapped between forces in the North and South, both intent on their speedy destruction.

This destruction started with a furious two-hour artillery bombardment on September 16th, and the American gunners were in their element. MacArthur had compared the men in the Pusan perimeter to 'beef cattle in the slaughterhouse', and now it was their turn. With a cold resentful fury American artillerymen went to work: muzzles recoiled and roared, gunners rammed shells into the open breeches as fast as they could pick them up, and shouted 'That's for Taejon' as they slammed the breech-blocks home. During that pounding – two hours in which it seemed that nothing could live – the sky over the North Korean positions was filled with the snarling whine of shells. As the barrage lifted, the men of the 8th Army advanced. And this time, as Walton Walker had grimly explained to all his regimental commanders, they would continue to advance: it was to be the end of stalemate – now it was to be a war of dash and movement; it was to be the end of 'bugging-out' and retreating.

So far the men of Aubrey Coad's 27 Brigade had been only bit-players in the main drama, although they had fulfilled their roles very much more than adequately. On September 21st they were elevated to star status.

Coad had his orders from Major-General John Church, commanding the United States 24th Division. The 1st Cavalry Division and the 5th Regimental Combat Team were to advance up the road towards Taejon and Seoul – the same road down which they had carried out a most depressing and punishing withdrawal. Coad admitted to a feeling of pessimism about this planned offensive: the 1st Cavalry Division had been fighting hard for ten days and had been gradually and progressively pushed back every single day; they were battered and sleepless. The 5th Regimental Combat Team, however, were fresh troops and anxious to show their mettle. The 24th Division, probably the most battered American formation of all, had been made up to strength by the integration of South Koreans, and were in reserve.

The role of 27 Brigade was to cross the Naktong River and protect the left flank. Then, when the American advance was well under way, they were to clear this flank for the main thrust along the road to Seoul.

The initial advance on the 16th could have done nothing but sadden any commander: the North Koreans, while Seoul was falling in the north, bowed to the inevitable and started a general withdrawal which loosened their defensive line as it ran north and east around thc perimeter.

The Americans, anticipating an abrupt collapse of Communist resistance, were made to pay dearly for every bitterly contested yard they advanced: far from crumbling, the North Koreans fought a succession of determined delaying actions to permit the orderly withdrawal of their main forces.

It seemed that Coad's first pessimism (which he was at pains to keep to himself) was justified: on the first day of the break-out the progress made was represented by an advance of 1,000 yards by the 5th Regimental Combat Team. Otherwise, as an official American report glumly phrased it, 'The battle for the hills had merely gone on into another day.'

For Walker the events of the 16th must have been a bitter disappointment. He had purposely delayed the break-out by one day so that all units could be told about the Inchon invasion; he hoped that this heartening news would infuse a new spirit into his tired troops.

But the truth of the matter was that the 8th Army as a whole was exhausted from the strain of the fighting since July. There comes a point when the best of divisions are battle-drunk; when they can slog on mechanically but lack fire, and so it was with the 1st Cavalry, the 24th, and the 25th: they had been through the heartbreaking retreat from the North and been attacked and pounded ceaselessly in the Pusan perimeter.

It was not until Walker committed his reserve, the 24th Division, to the battle that the offensive started to roll, and once it gathered momentum there was no stopping it.

[4]

In his dispatches about the early doings of 27 Brigade, Aubrey Coad wrote: 'I want it to be clearly understood that the 1st Argyll and Sutherland Highlanders and the 1st Middlesex were perfectly normal battalions with quite a large number of National Service soldiers in them.' He meant nothing disparaging about this; he only wanted to make it clear that the first British soldiers to fight in Korea carried no superman tags: they were not the Brigade of Guards, the United States Marines, the Commandos, the Paratroops, the SS, *Les Paras*, or any other formation associated with almost superhuman qualities of toughness and courage. He only produced this slightly deprecatory explanation once: after the 21st he had no need to . . .

The conduct of the Argylls and Middlesex in the crossing of the Naktong won the wholehearted admiration of everyone who witnessed it. Brigadier Aubrey Coad, whose emotions were always under severe control, confessed that he 'gulped with pride' as he watched his men crossing the Naktong. For not a single man faltered on his way across.

It was, as an American officer conventionally described it, 'a rugged assignment': the only way across was a precarious and rickety bridge, which had been hastily erected by American engineers. The only formation possible was in single file. The bridge was under continuous fire from a roving self-propelled gun (referred to as as 'the bastard' by everyone who was there) and mortars. Mercifully the steadiness of 27 Brigade and the inaccuracy of the shooting ensured that the crossing was an unqualified success.

When the 1st Middlesex went across the Naktong as steadily as if on an Aldershot parade ground, Lieutenant-Colonel Andrew Man was at their head – it never for one second occurred to him to be anywhere else. This was not ostentation on Man's part, and he sought no glory: by his uncomplicated reasoning, if his battalion was going to cross the river then the commanding officer should cross it first.

'Take the high ground,' exhorted the instructors at West

Point. 'That's a swell piece of real estate an' I want it bad,' said American generals of hill features in the Korean War. There was high ground and 'real estate' in plenty to the right and left of the road leading to Songju, and General John Church wanted it all. These Limeys were going to get it for him.

Church was a man who gave an order, and expected it to be carried out without throwing in arguments, objections, provisos, or governing clauses. Coad's laconic 'We'll do it, sir,' was music to his ears and all he wanted to hear.

Coad's orders to Man and Neilson were brief and to the point: the Middlesex were first to secure a hill feature known, because of its peculiar shape, as 'Plum Pudding', and another hill – sheer, daunting, and 900 feet tall – which later became known as 'Middlesex Hill'. The Argylls would go for Hill 282 on the left of the road. Coad knew that the eyes of the world would be upon 27 Brigade, but he did not consider it necessary to tell Man and Neilson this: they knew it well enough . . .

Second World War veterans of the 1st Middlesex – they included a small nucleus of officers and other ranks who had fought in North Africa, Italy, and France – affirmed that men of the regiment had never attacked with greater courage and offensive spirit than on September 22nd, 1950. Platoon sergeants, who had initially regarded National Service soldiers with a measure of suspicion, took it all back unreservedly.

As dawn broke that morning, No 4 Platoon of 'B' Company, 1st Battalion the Middlesex Regiment, stretched their cramped limbs and thought about the forthcoming attack and breakfast – in that order. Above them in the gloom loomed 'Plum Pudding' Hill, their objective: it looked menacing and inhospitable. Soon they would be going up that hill in the same way that they had practised platoon attacks on exactly similar looking hills in Hong Kong. But there were North Koreans on Plum Pudding: determined, securely entrenched, and intent on killing every last one of them.

Second-Lieutenant Christopher Lawrence strode among his men like a genial young Napoleon: he noticed with approval that Private 'Greg' Peck, having temporarily jettisoned the hated bazooka, had stopped grousing; he caught the latest 'smoking-room' story from Corporal Joe Pentony's repertoire; Sergeant Butters, who had fought in North Africa and landed on D-Day, was telling the men that there was nothing to worry about – this job, he declared, would be a piece of cake; Private Sam Buckle and 'Chalky' White were telling one another that it would be for the good of the community if the other was killed; Private Ridge remembered that Peck still owed him a dollar from Hong Kong days.

There were two American tanks with Lawrence. The officer in charge, a cheerful young man who, defying dress regulations, wore a long, peaked, red baseball cap, said: 'We'll be in there slugging with you, mac.' Then, as if to illustrate Anglo-American solidarity, he suddenly fired a long burst of machine-gun fire for no attributable reason at the top of 'Plum Pudding'.

Lawrence looked at his watch. The second hand was creeping towards seven o'clock. He said: 'Right, chaps – time to go.'

The attack on Plum Pudding Hill was in the words of Andrew Man, 'a most gallant affair': the average age of the platoon was just 20, yet they scaled that hill like veterans. Christopher Lawrence led the assault in much the same way as his father had led a platoon of the regiment in 1916 – well in advance of his men and cheering them on.

No 4 Platoon followed Lawrence up every bitterly contested foot: Corporal Pentony dispatched four North Koreans with his sten gun, before falling to a burst of machine-gun fire; Butters, Peck, Buckle, White, and Ridge – they were all there: picking off fleeing North Koreans with rifle shots; grenading and bayoneting them from their trenches.

There were brave North Koreans in plenty holding Plum Pudding, but they could do little against the disciplined impetus of No 4 Platoon's bayonet charge, and when it came to cold steel they had clearly met their match. There were a few agonized squeals, such as made by a trapped rabbit, as the bayonets went home. Then they broke and ran to the top of the hill and down the other side, pursued by the bullets and obscenities, which come easily enough to the lips of Middlesex men at any time, but still more so when the bayonets are fixed and the blood is at white heat.

Andrew Man described the spectacle of Christopher Lawrence on Plum Pudding Hill after the attack as 'one of the most heartening sights of the entire campaign'.

Meanwhile, 'A' and 'D' Companies, supported by the machine-gun platoon and American tanks, had secured 'Middlesex Hill', and here too it was clear that the North Koreans were making no last ditch stands. They fled down the reverse slope, and their progress was watched with fascinated enjoyment by an American artillery observation officer chewing on a mutilated cigar.

'Boy,' he exclaimed, 'ain't that a honey of a target!' He barked instructions into a field telephone, and soon shells were falling among the fleeing enemy. The tank officer wearing the red baseball cap, not to be outdone, loosed off long bursts of machine-gun fire.

On reading the terse words of the Intelligence Summary at 27 Brigade Headquarters the attack did not appear to amount to much: 'Plum Pudding Hill and Middlesex Hill secured. Estimated enemy casualties 50. Own casualties 1 officer 6 other ranks killed; 7 other ranks wounded.'

Nor was it much to read about in the English newspapers. In England morose breakfasters remarked that there was 'nothing in the papers as usual' and resumed the national pastime of griping about the cost of living.

But Second-Lieutenant George White would now never join the Royal Army Ordnance Corps. White was in the Middlesex Regiment, but not of it: the War Office had decreed that young officers destined for the R A O C would serve an attachment with an infantry battalion after they left Sandhurst. George White, whose military future would normally have been bounded by stores, indents, and forms in triplicate, was killed in the attack on Middlesex Hill.

The Roll of Honour of the R A O C is a long one, but young second-lieutenants with six months' service were rarely to be found scaling precipitous hills under fire at the head of a platoon of infantry. George White did that and more: he led his platoon with enormous dash and gallantry, and at the moment of victory, while recklessly exposing himself to enemy fire, he was killed. He was posthumously awarded the American Silver Star.

Corporal Joe Pentony, with six bullets in his stomach, was carried back on a stretcher. He displayed, according to Christopher Lawrence, 'quite extraordinary courage': he borrowed cigarettes from the stretcher bearers and assured every one in earshot that he would be back inside a week. In hospital his repertoire of stories was as inexhaustible as ever. But on September 26th, Corporal Joe Pentony died.

Privates Bristow, Hayfield, Russell, Sharpe, Tredget, and West had consumed their last 'San Mig' beer, scrounged their last fag, written their last letter home, pestered the CSM for a weekend pass for the last time.

It wasn't much, but it wasn't nothing . . .

[5]

Every year Argyll and Sutherland Highlanders meet at regimental reunions to re-live the battles of bygone days. A few are old, rheumy, and white haired: the majority are in the greying and balding wilderness between 45 and 60.

Now a new generation of Argylls with terrible tales to tell has sprung up. To the vintage Jocks they are 'weans', scarcely dry behind the ears. But they too have their day to remember – a proud day in the regiment's history, but one overlaid with fearful tragedy.

The day is September 23rd, 1950. And the place is a hill in Korea – Hill 282 . . .

The Middlesex were firmly on their positions, and now it was the turn of the Argylls. Colonel Neilson's orders were that the assault on Hill 282 would be carried out by Major Alastair Gordon-Ingram's 'B' Company and Major Jim Gillies's 'C' Company. Major David Wilson's 'A' Company was to occupy a subsidiary hill feature from which they would cover the main assault.

'A' Company had an easy job, although at first sight Wilson's task looked daunting enough. North Koreans were clearly visible on top of the hill, in positions from which they could machine-gun the Argylls with insulting ease. Not that this would have

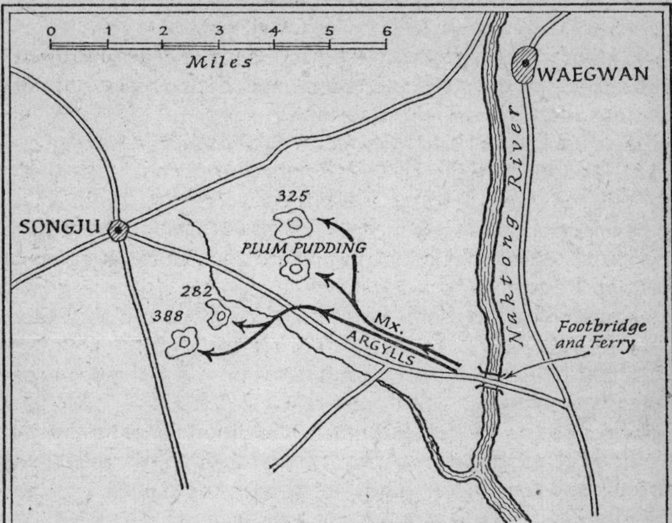

MAP B THE ACTION AT SONGJU

© CASSELL & CO LTD 1967

deterred Wilson and his men in any way: their orders were to get on to that hill and they were going to do so, come hell and high water.

As David Wilson halted his men and dubiously surveyed the hill – he would take it all right, but casualties were certain to be severe – three American tanks came at a leisurely speed round the corner. Wilson waved to them to halt.

They were, it turned out, part of a Reconnaissance Company of the 24th United States Division. What strange freak of fate had brought them there at that particular moment, Wilson neither knew nor cared. He only knew that their presence would probably save fifty casualties.

The tank commander was a grim-looking Master Sergeant who seemed to have emerged straight from Hollywood: he chewed on a damp cigar and sported a three-day growth of black stubble. His driver, a beardless young man who was chewing on a piece of gum with dogged perseverance, said nervously: 'Hell, sarge, what we stoppin' for?'

'Button it up,' said the Master Sergeant tersely. He favoured Wilson with a melodramatic salute from the turret of his tank. 'Something I can do for you, Major?'

'A small favour,' said Wilson.

'Just name it, Major.'

'You see that hill?'

'Check. On top of which I see gooks also' – the Master Sergeant's speech, like his appearance, came from the pages of Damon Runyon.

'Exactly,' said Wilson. 'I wonder if you'd care to shift them for us.'

'Roger,' said the Master Sergeant. 'Man, you just got yourself some tank support.'

The ensuing expenditure of ammunition was prodigious, even by United States Army standards, but the effect was drastic and immediate. The North Koreans melted from the position.

'Nice job,' commented Wilson. 'Thank you very much.'

'Any time at all,' said the Master Sergeant: 'Just ask for Joe and we'll come running.' He produced his peculiar salute once more, barked an order, re-lit his cigar, and the tanks roared round the next bend.

'A' Company walked unopposed on to the position and dug in, having suffered no casualties. The opening phase of the Argylls' attack had been neatly and quickly concluded, and protection ensured for the start-line of the main assault.

Meanwhile, 'B' and 'C' Companies were experiencing the most uncomfortable and nerve-racking experience in the infantryman's book – the uneasy 'lying-up' period before a dawn attack: a period in which there is precious little sleep, no talking and no smoking. The attack was scheduled for 5.20 AM on September 23rd.

Two platoons of 'B' Company, commanded by Lieutenant Peter Mackellar and Sergeant John O'Sullivan were the first to go into the attack. They had a difficult scramble through the dense fir trees and loose rocks that covered the almost sheer slope. But they were trained to scale such slopes as these – it was no worse than a dozen such hills in the New Territories – and in just under an hour of climbing, in a silence broken only by heavy breathing, they were less than fifty yards from the enemy position.

They paused to get their breath, and during this brief rest they smelled a smell which must be one of the worst that can assail a man's nostrils – the odour of fish being cooked, Korean or Japanese style: it is an odour so vile that patrols in Burma, on sniffing it, have retched loudly, thus giving their positions away. But no Argylls were sick that morning . . .

The condemned men did not eat a hearty breakfast; indeed, they ate no breakfast at all. A few reached their weapons and loosed off despairing shots, and a sentry got in a long burst with a burp gun which wounded both Mackellar and O'Sullivan as they charged over the crest of the hill. It stopped them, but did nothing to quench their offensive spirit: as they lay on the ground, they continued to shout encouragement at the Jocks who surged past them.

Corporal Robert Sweeney, an unimaginative soldier, had never visualized himself commanding any formation bigger than a section of ten men. But now, with both leaders down, he was (as they say in National Hunt racing) left out in front. Yelling like a dervish, he dispatched the burp gunner with a burst from his sten gun; behind Sweeney came Lance-Corporal Ron Syme, 'Sailor' Kelly, Kinnear, McEwan, 'Patsy' Quinn, 'Rob' Roy . . . with grenades, bayonets, and rifle butts they went through those North Koreans like a knife going through butter.

It was a true Highlanders' charge of the type that delights battle historians: fifteen enemy dead were counted on the position, and at least a dozen more were seen to fall on their headlong flight down the reverse slope. The North Koreans on 282 had shown themselves to be as allergic to cold steel as had those on Plum Pudding.

Hard on the heels of 'B' Company came 'C'. By 6.18 AM, at the cost of one killed and six wounded, Alastair Gordon-Ingram was able to report to Colonel Neilson that Hill 282 was in Argyll hands.

On all sides men were digging feverishly in preparation for the inevitable counter-attack. It was not slow in coming.

It started with shelling and mortaring, which killed four men and wounded nine more. Then Second-Lieutenant Jock Edington, in one of the more isolated platoon positions, reported enemy infiltration on his left front. Majors Gillies and Gordon-Ingram knew that North Korean infantry would soon come at them in force to wrest back the ground they had lost.

Experienced soldiers both, they were not at first unduly perturbed: shrewdly delivered artillery fire should effectively break up any counter-attack and the Jocks with their light automatics, rifles, bayonets, grenades, and guts would do the rest. 'B' and 'C' Companies had with them American artillery forward observation officers – two bustling and energetic young men who, in spite of their seemingly interminable conversations on field telephones, clearly longed to bring guns to bear on the enemy.

At about 8.45 AM, when Edington was reporting still larger numbers of North Koreans forming up for the attack, the Argylls on Hill 282 suffered their first cruel disappointment.

A crestfallen and infinitely apologetic American artillery officer approached Major Gillies at 'C' Company Headquarters. He wished Gillies to understand clearly that it was none of his doing: if he had his way, he would stay and ensure that the guns blasted those goddamn gooks into little pieces. But he had had his orders from regimental headquarters – he'd got to go back. Personally, he felt as mean as hell about it, but there it was. If he didn't obey the order then he'd get the book thrown at him.

'You mean,' said Gillies, scarcely able to believe his ears, 'that we're to operate without any artillery support whatsoever?'

The artillery officer shrugged helplessly. 'I'm afraid that's about it, Major.'

Gordon-Ingram and Gillies protested volubly to Colonel Neilson who in turn protested to Brigadier Coad. Coad's protest was the loudest of all, and reached the ears of General Walker himself, but at the time nothing could be done about it. Leslie Neilson was in the unpleasant predicament of seeing his men isolated in face of the enemy without being able to do anything to help them.

'B' and 'C' Companies were out on a limb . . .

As Gillies and Gordon-Ingram had foreseen, the North Koreans reacted vigorously and counter-attacked with vicious intensity, There was increased shelling and mortaring and a dozen more Jocks were killed or wounded.

To the infantryman crouching in his foxhole there is no more comforting sound than the crump of artillery shells falling on the enemy in front of him. But 'B' and 'C' Companies were denied this luxury, and both company commanders swore loud and long at the decision to withdraw artillery support. 'It's a great shame,' observed Gordon-Ingram sourly, 'that the genius who gave that order isn't up here.'

Following the shelling and mortaring came the North Korean infantry: the very thick vegetation made it easy for them to approach to within a few yards of the Argylls' position. They were able to place machine guns in such a way that they could pour a steady stream of accurate fire into the defenders. At first the main brunt fell on 'B' Company, and in the space of less than half an hour No 4 Platoon was reduced to seven effective fighting men. Edington's platoon was in imminent danger of being complerely overrun and withdrew into the main perimeter.

The situation was worsening every minute. But although ammunition was running dangerously low and casualties were mounting, the Argylls were giving as good as they got – an encouraging number of North Koreans were falling to every burst of bren-gun fire – and both Gillies and Gordon-Ingram were confident that the position would be held. It had to be in any event, because the order they had received had been that good old one which the British Army loves to trot out on every possible and impossible occasion – 'the position will be held to the last man and the last round'. The Argylls on Hill 282 had not questioned that order, and they were not going to.

The most pressing problem of all was the evacuation of the

wounded. There was only one way to safety for them: down the 900-foot-high hill. Men who were desperately needed in the firing line assisted the wounded to the bottom of the slope, but the round trip took at least one hour – an hour in which the North Koreans drew ever closer and still more Argylls were wounded.

Ensuring that no man took longer than necessary on his errand of mercy was Company Sergeant-Major Tom Collet, who grimly informed them that he was timing each journey with a stop-watch. 'No stopping at the bottom of the hill and lighting a fag, got it?' was his executive word of command, and no man did, for the CSM could be terrible in his wrath and the Jocks of 'C' Company feared it at least as much as the bullets on the position.

Tom Collet was the almost perfect sergeant-major of fiction with a shattering parade-ground voice and a cold eye for defaulters: today he keeps a pub in Suffolk and the authoritative air lingers on, as obstreperous customers have discovered. He was always impeccably shaved and his sparse, sandy hair was cut once a week. On parade his eyes were like drops of ice, and CSM Tom Collet was nearly, but not quite, always on parade. He was an Argyll to the bone and had been since he had enlisted as a 14-year-old drummer boy seventeen years before. Major Gillies said afterwards that he made up for the lack of artillery support.

During the battle Collet was a tower of strength: wherever the fighting was thickest, and that was almost everywhere, he was to be seen and heard – encouraging, joking, cajoling, deriding; he fired a bren gun, hurled grenades, and did deadly work with rifle and bayonet.

Later in the battle he made himself responsible for the evacuation of casualties. Stern and unbending disciplinarian that he was, wounded Jocks soon discovered that he could minister to hurt and helpless men with a mother's tenderness. A non-smoker himself, he seemed to have an endless supply of cigarettes; for men parched with thirst he had an ever-ready water bottle; he applied field dressings with dedicated and loving care. Wounded Jocks, who had hitherto only encountered the rough side of Collet's tongue, discovered that he seemed to be motivated by an almost divine mission – to succour the wounded without thought for his own safety.

The situation was eased by the appearance of the battalion stretcher bearers – in battle this unglamorous but essential task was performed by the pipers and drummers. As pleased to see them as anyone was CSM Collet; now he could get back to his proper job, which was gingering up 'C' Company – even now he could see that idle, horrible man Private Kelly lighting a fag when he should have been bringing up ammunition . . .

With the stretcher bearers came the second-in-command of the Battalion, Major Kenny Muir – one of the greatest Argylls of them all.

Like Tom Collet, Muir was an Argyll to the bone. Aged 38, he was unmarried because the Argyll and Sutherland Highlanders – his father had served in the regiment before him – was his first and only love.

Some people have the idea that the second-in-command's function in battle is merely to wait for the Commanding Officer to be killed or wounded. There is considerably more to it than that: he is, in principle, 'LOB' (Left out of battle) and occupies himself with transport, ammunition supply, communications, and all other matters which come under the heading of 'administration'. He does not, as a rule, concern himself with casualty evacuation. But on the morning of September 23rd Kenny Muir was in agreement with Marshal Foch, who said: 'To hell with principles! What's the problem?'

The problems, Kenny Muir discovered, were numerous and becoming more so every minute. These he communicated to Colonel Neilson, who demanded with some asperity what the hell he was doing there.

'Just keeping in touch, sir,' said Muir disarmingly. Neilson let it go at that: if the situation on Hill 282 was as bad as it sounded, then there was no man he trusted more implicitly than Kenny Muir.

'B' and 'C' Companies had become inextricably mixed, and Muir decided to reorganize the companies into one force under his command: he made fresh dispositions: he tightened the perimeter; he centralized the collection of ammunition from casualties and its redistribution to the firing line.

The dynamic personality of Kenny Muir was soon impressed upon every officer and man of his small and steadily dwindling command. The spirit of the leader became the spirit of the men. And the spirit of Muir was one, not so much of hope, but of certainty of ultimate victory – one of calm and confident self-reliance and firm belief in the worth of his men. Every Jock felt that the eye of Muir was upon him; that he was at his side; that he knew not only what the soldier did but what he thought.

With such a man as Kenny Muir to command, lead, and discipline them, watch over them and fight for and with them, the Argylls knew that they could not and would not be vanquished.

They were being attacked frontally and from both flanks. Nevertheless, in the opinion of Gillies and Gordon-Ingram, Muir had so warmed up the defence that the situation was in hand and the hill would be held; indeed, the enemy machine-gun fire was slackening and the assaulting infantrymen were showing signs of having had enough.

The situation, however, was still serious: there was a large force of North Koreans on a ridge some 1,500 yards away to the left of the Argylls' position. As artillery and tank support were unobtainable, Muir spoke on the radio to Colonel Neilson requesting an air strike on that ridge; although it could not hit those North Koreans actually attacking, it could strike a crippling blow at reserves moving up to the attack.

The news that United States Mustangs were on the way soon spread to every man in 'B' and 'C' Company. 'That'll fix 'em,' said C S M Collet with satisfaction; 'but it doesn't mean that you lot can sit on your arses looking at the sky. Watch your front.'

Recognition panels were put out so that the aircraft could distinguish the Argylls' position. Then at about 12 noon three aircraft circled over the position: Jocks stood up and cheered, waving their arms in greeting.

And then it happened. Each Mustang in turn made a low 'run-in' and sprayed Hill 282 with machine-gun bullets, following this murderous assault with napalm bombs. The whole fearful tragedy was over in two minutes – two minutes in which Hill 282 became a veritable sea of fire from end to end.

Amid this dreadful maelstrom men were struck down by machine-gun bullets, almost before the welcoming cheers were out of their mouths; others, directly in the path of the flames from the napalm bombs, were caught by the fire and flayed and roasted from head to foot; men already wounded, and powerless to move out of the way of the licking flames, were grilled as thoroughly as any steak where they lay. There were not only bullets coming from the sky: the reserve ammunition supply was caught by the flames and exploded in every direction. Hill 282 presented an authentic picture of hell.

It would have been a hideous experience even for battle-hardened troops. The Argylls did not abandon the hill – they were literally blasted and scorched off it.

About thirty men made their escape from the inferno of the hilltop – men wounded, men burned, men who were miraculously neither, men shocked out of coherent thought. But this was no panic-stricken flight: they had left the hill because nothing could live on it.

They huddled on the ridge below the hill. When the first pulverizing shock started to wear off, they looked unhappy and a little shamefaced, for there was no gainsaying the fact that ground had been lost – and the Argylls have always hated giving up ground to the enemy. They would have continued to hold it with rifle butts and bayonets and boots and fists, but a disaster as shattering as this was something beyond their ken. They were still belligerent, but disorganized. They needed a leader, and they needed him quickly.

They still had one. Kenny Muir had left the hill together with Captain John Penman, Major Gordon-Ingram and CSM Murray of 'B' Company. Undoubtedly a complete withdrawal would have been justified at this time: indeed, Colonel Neilson and Major Wilson, who had been horrified and impotent spectators, marvelled that any Argylls remained.

Muir's mind, however, was still on the battle. He observed that the enemy had not taken immediate advantage of the situation and had not occupied the top of the hill. He noticed something else: from Hill 282 he could still hear the defiant clatter of a British bren gun and rifles. Clearly a small party of Argylls were still holding out. Therefore, reasoned Muir, the hill must be retaken and held so that the wounded (assuming that any men had lived through the devastating air strike) could be evacuated.

The fire on the hill came from a party of five Jocks, commanded by Private William Watts. They had escaped the holocaust of the air attack and seen men running from the hill. It seemed to Watts to be a reasonable course of action: bullets were flying everywhere and the hill was on fire – but no one had told *him* to leave, and he wasn't going to while he had his bren gun and two full magazines.

To his own amazement Watts found that he had qualities of leadership, for which his officers and sergeants had previously searched in vain. His voice brooked no disobedience; to the others it seemed that, if they had to die, Bill Watts was a good man to die with.

Muir, Gordon-Ingram and Gillies rounded up all men fit to fight – some thirty in all. They were going back, Muir informed them grimly: the North Koreans were not going to be permitted to occupy the hill which they, the Argylls, rightfully owned. The heart had gone out of some of the men, who had been so murderously ousted from the position, but Kenny Muir put it back. Ordering Major Gillies to give them covering fire, Muir said to the men: 'Right, we're going in,' and at the steady loping trot of the well-trained soldier these thirty men went towards the by now smouldering hilltop.

The Argylls had never faltered since they first set foot in Korea, and they did not do so now. With Muir at their head, waving his Balmoral bonnet and cheering them on, they went up that hill in the face of scything fire.

By the time they reached the crest there were only fourteen men still on their feet, but having recovered the position Kenny Muir was determined that it should be held and he imbued the survivors with that same resolve. Ammunition for his sten gun was finished but he seized a two-inch mortar, and with Alastair

Gordon-Ingram acting as loader, fired it at close range with deadly effect. He was in the act of firing the mortar when he was hit by two machine-gun bursts in the stomach and thigh. Even as he was being carried down the hill he continued to shout encouragement to his men, and when he could no longer shout he whispered and when he could no longer whisper he died. His last words were, 'The gooks will never drive the Argylls off this hill.'

Kenny Muir was posthumously awarded the Victoria Cross. The official citation had it thus: 'Major Muir's actions were beyond all possible praise . . . the effect of his splendid leadership on the men was nothing short of amazing and it was entirely due to his magnificent courage and example and the spirit which he imbued in those about him that all wounded were evacuated from the hill, and, as was subsequently discovered, very heavy casualties inflicted on the enemy in the defence of the crest.'

The gallant few who remained had become a spent force: Gordon-Ingram could only count ten whole men in 'B' Company, and 'C' were in little better case.

'The position will be held to the last man and the last round . . .' – they would soon be reduced to the last man and the last round was very nearly shot. Gillies and Gordon-Ingram agreed that one last head-on assault from the North Koreans would wipe them out completely. They informed the Commanding Officer of this and Leslie Neilson replied: 'Damned well done. Now get out of it.' It was time to go . . .

They came down from Hill 282, which now looked like a dirty corner of the Somme in 1916 with its blasted trees and thousands of empty cartridge cases; it also smelled like the Somme – the sickly, cloying smell of death hung over the whole area like a pall.

They came in a sort of cosmic slow motion: their faces had a greyish tinge, their eyes wandered, their mouths drooped open. Looking into the faces of men coming out of battle it is their jaws you notice first – they seem to hang loose, for they have been clenched for hours and relax first of all.

But not a man looked beaten, any more than any survivor of the old 2nd Battalion on the Slim River had looked beaten.

Grossly outnumbered and without support of any description, they had fought the North Koreans to a standstill for nine hours and well knew who were the better men. They had enriched with honour and glory the traditions of their regiment which had always consistently lived up to its motto, '*Sans Peur*'.

The Argylls' casualties had been grievous: in addition to Kenny Muir, Second-Lieutenant David Buchanan, cousin of Neil who was lost on that first ill-fated patrol, was killed in the air strike. Up to the time of his death, he had commanded his platoon with enormous gallantry in one of the most hard-pressed sectors of the position. Casualties totalled 17 killed and 76 wounded – virtually the strength of an entire company.

There were several men of the Middlesex helping the wounded. Horrified witnesses of the air strike, they had dashed without orders to the help of the Jocks. Between the Jocks and Midds existed a fast friendship, and several Middlesex officers had to shout 'Stand fast!' in the H M S *Birkenhead* manner to stop many other would-be helpers taking off for Hill 282. Strictly according to the book, they were guilty of abandoning their positions, but there was not an officer or man of either battalion who did not give them honour for it.

Here came two fit men; here two more, supporting a man with a foot partially severed; a man with a head wound, roughly bandaged; a man trouserless, face down on a stretcher, with deep and cruel first-degree burns on his back and buttocks. Here came still more wounded: Lance-Corporal Harry Ward, who had recently held the cosy sinecure of Battalion post corporal: he had been hit in the arm by a burst of machine-gun fire, and this arm was subsequently amputated; Private Vic Kinnear with one shattered leg, Sergeant O'Sullivan with two; the red-haired Sidey; Kelly, Mitchell, and Quinn; two 'Scots by adoption' Irishmen – Docherty and Delaney; Corporal Sweeney, who had led the first charge, and was to win the Military Medal for it; the indomitable Watts, whose refusal to leave the hill was to earn him the American Bronze Star; men with names which were in the ranks of the 93rd at Balaclava: McNaughton, Campbell, McEwan, Roy, Dunbar, McPherson, McDonald, McLaughlin . . .

Came four wounded officers: Captain John Penman (he had

won the Military Cross in the Second World War, and was shortly to add a bar to it); Jock Edington, Peter Mackellar, Jim Stirling – these last three represented a hundred per cent casualty rate in 'B' Company platoon commanders.

Here came C S M Tom Collet, and it seemed inevitable that he should be last down and carrying a wounded man. His burden was Lance-Corporal Syme, who expressed the incongruity of his position by saying: 'Blimey, never thought I'd have my arm round *your* neck,' to which Collet grimly gave answer: 'It's going to cost you a pint' – and it subsequently did . . .

How such such a disaster could have happened in spite of the display of recognition panels remains unexplained. The remorse of the American Air Force was expressed not only in words but in deeds, and a generous cheque was subscribed by officers and men of their 93rd Bombardment Wing for the families of those killed and wounded. In acknowledgement of this Lieutenant-General Sir Gordon MacMillan, the Colonel of the Argyll and Sutherland Highlanders, expressed the correct view of the whole tragic episode: 'The Regiment's friendship with the United States Air Force can never be impaired by having suffered on one occasion from the risks which are inseparable from operations in modern war. Every report I have received from the Battalion . . . has spoken in glowing terms of wonderful cooperation, and no hard feelings must arise as a result of this incident.'

The Argylls bore no malice about the tragic affair on Hill 282, and this attitude of mind was best expressed by Private Peter Sinclair from a stretcher. 'It couldn't be helped,' he said – and no Public Relations Officer or politician prompted him in this speech – 'it was just one of those things.'

SIX

THE scene at Pusan docks on September 27th, 1950, was much the same as it had been when the Argylls and Middlesex had arrived: the same Korean brass band murdered the works of Sousa; the same American Negro band with the same gigantic drum-major at their head played *St Louis Blues*; there was the girls' choir, the bouquets, and the local dignitaries. The welcoming committee at Pusan, American and South Korean, had their formula and knew a good thing when they saw it.

The air of Pusan was as foul as usual, but on this day it was blue as well – blue with appalling epithets, frightful oaths, hideous blasphemies, unmentionable comparisons. The voices were strident with a broad twang and sharpened vowel sounds: they were voices immediately, and in some quarters apprehensively, identifiable as Australian.

The men streaming down the gangways of the American ship *Aitken Victory* seemed to be bigger and tougher looking than the men of 27 Brigade; their faces were square, leathery and tanned, and on every one was a broad grin. They were certainly older, and on a large number of left breasts could be seen ribbons, outward and visible signs of war service in North Africa and the Pacific. They wore slouch hats at rakish angles. The Americans on the dockside had referred to the Argylls and Middlesex as a 'rugged-looking bunch'; with the appearance of these newcomers to Korea they said it again with even greater feeling.

This, then, was the 3rd Battalion The Royal Australian Regiment. They had come to add a very considerable contribution to the Korean War.

The outbreak of the war in Korea on Sunday, June 25th, 1950, did not initially make a big impact in Australia; indeed, it did not even appear in the Sunday papers at all. Australia had trouble enough of her own, for on the northern coast of New South Wales there had been devastating floods: rivers swollen by cyclonic storms had driven thousands of people from their homes and caused millions of pounds worth of damage. Australia in the last week of June had acute refugee problems of her own without getting involved in Korea.

The majority of Australians knew little and cared less about Korea.

On June 29th the Prime Minister of Australia, Mr Menzies, announced that two Royal Australian Navy ships in Japanese waters, the destroyer *Bataan* and the frigate *Shoalhaven*, had been placed at the disposal of the United Nations in support of the Republic of Korea. The Royal Australian Air Force 77 Squadron was alerted for immediate operations. Clearly Australia was getting into this war too, and Mr Menzies promised the United States House of Representatives the assistance of Australian troops of brigade strength. This force would be drawn from Australia's minute regular army, units on occupation duty in Japan and volunteers.

On July 26th the acting Prime Minister, Mr Arthur Fadden, broadcast the news that Australia would send troops to Korea. At 9 A M on July 27th recruiting offices for volunteers opened; by ten o'clock they were flooded. The regular battalions in Australia – the 1st Battalion The Royal Australian Regiment at Ingleburn, New South Wales and the 2nd Battalion The Royal Australian Regiment at Puckapunyal, Victoria – volunteered to a man for service in Korea. Not all could go, but the fighting spirit of Australia was proven beyond all doubt. The volunteers came from all over Australia – from the big cities like Sydney, Brisbane, Melbourne, Hobart, Perth, and Darwin and from dusty, parched little towns in the Western outback: there were bank clerks and barmen; huge, bronzed lifeguards from the beaches; sheepmen, cattlemen, and timber fellers.

There were men who had worked the tin mines in Tasmania and the opal mines at Andamooka and men who had humped wheat at the outback sidings in New South Wales and men who had tried their luck in the Mildura vineyards. There were men who had seen the inside of Long Bay Gaol and others who were destined to. There were drifters, swagmen, and bushwhackers. There was also a high percentage of Englishmen who, finding that Great Britain in 1945 did not come up to expectations, emigrated and announced on arrival in Australia that 'the old country was played out, mate'. If they said this with the requisite force Australia accepted them with alacrity. Many of these men had travelled incredible distances to join up; the record for endurance was held by Private Robert Hellyer, a 20-year-old cattleman, who hitch-hiked 1,300 miles to Adelaide. They volunteered for a bewildering variety of reasons, but basically because they wanted to get into a fight – any fight. Most of them knew all about the unpleasantness of war, and had impressive displays of medal ribbons to prove it: they had fought the Germans in the Western Desert, Greece, and Crete and the Japanese in New Guinea, Borneo, Bougainville, Papua, New Britain, Wewak, and Malaya. It didn't particularly matter to them who these jokers in Korea were, but anything the Yanks and Pommies could do they could do a bloody sight better.

Many of them lied barefacedly about their ages: the youngest volunteer, aged 16½, convinced a recruiting sergeant that he was 18; the oldest, who had fought in New Guinea, admitted under close questioning to 38, whereas he was in fact 46. Birth certificate? Don't you come the raw prawn with me, mate . . .

These older men were the 'Ratty Diggers', so called by some of Australia's young men because it was thought that having survived one major war they would have sufficient sense to avoid another.

At least half of the Aussies who disembarked at Pusan on September 27th were 'Ratty Diggers'. But the day was not far off when Brigadier Aubrey Coad was to call the 3rd Battalion The Royal Australian Regiment 'the finest fighting battalion he had ever seen'. And he meant it.

Overjoyed as he was to have a third battalion added to his 'Cinderella' brigade, Coad initially confessed to certain misgivings when he heard that they were to be Australians: he had heard that they did not like taking orders from anyone, least of all from a 'Pommy'. He had heard, too, alarming tales of extraordinary immoderation in Cairo during the war – 'marvellous chaps in the line,' a friend had told him, 'but put 'em in a big city with a handful of back pay and God help you.' He knew that platoon commanders were invariably known by their christian names to their men, and that more senior officers were sometimes accorded the formal appellation of 'skipper'. He had heard that their touchy discipline sometimes disappeared altogether and that only their own officers could do anything with them. They were, it was darkly hinted, over-fond of drink and women: in Pusan, at least this would present no problem – even the water was barely drinkable, and the appearance of the women would daunt even the toughest and most lecherous 'Ratty Digger'.

But Coad had also heard of their amazing gallantry in action; he knew about the Victoria Crosses won in two World Wars – usually crazily impossible one-man acts of heroism with rifle and bayonet or bren gun fired from the hip.

Coad need not have worried, and after meeting the 3rd RAR at Taegu on September 30th, he promptly ceased to do so. Coad visited the 3rd RAR at Taegu, and immediately liked what he saw; this was mutual. He found that the Commanding Officer, Lieutenant-Colonel Charles Green, was just 30 years old; that he moved with the natural grace of an athlete; that he radiated enthusiasm and confidence. Green was a dedicated soldier, who had learned his trade the hard way in the jungles of New Guinea. Coad realized that here was a fighting battalion which wanted to go into action at the earliest possible moment: for them no preliminary shakedown period for 'green' troops would be necessary – no gradual introduction to war in a quiet corner of the line, even if there had been such a thing.

Coad discovered, among other things, that he was a 'grey-haired old bastard'. He had overheard a conversation of a group of soldiers – Australians never bother to lower their voices, even when generals are in their lines – and deduced, correctly, that they were referring to him. Aubrey Coad is a Commander of the Bath, a Commander of the British Empire, a Companion of the Distinguished Order (twice) and holds the American Silver Star and the Order of the Legion of Merit. However, he still values that Australian accolade, spoken by an anonymous private, as highly as any of these decorations.

[2]

By October 1st total collapse of North Korean resistance, south of Seoul was in sight. The United States 24th Division, against strenuous resistance, had taken Taejon on September 28th: they had entered it, as they had left it in July, in flames; the 1st Cavalry Division had made contact with the 7th Division near Osan; General Walton Walker's 8th Army had met the 8th

Corps, which had landed at Inchon and captured Seoul. On September 29th, with Seoul and the whole of South Korea in his hands, General MacArthur halted the offensive and regrouped his forces: he had achieved the first objective of the United Nations armies – 'to repel the armed attack'. It seemed that all that now remained was to achieve the second – 'to restore international peace and security in the area'.

But for Douglas MacArthur military success brought as its only reward political dilemma. MacArthur was faced with two choices: to establish a *cordon sanitaire* south of the 38th Parallel, where the Korean peninsula is at its narrowest (if he followed this course, the North Koreans might recover their losses, build up their strength once more and reiterate the threat of Communist aggression), or, at the risk of being labelled 'aggressors' themselves, for the United Nations to advance beyond the 38th Parallel and carry the war into the enemy's country.

It was a situation of enormous delicacy. There was a widespread desire to keep the Korean War within bounds. In any event, the object of the United Nations was to prevent or localize wars and not to turn little wars into big wars: Russia and China were very far from being disinterested in the Korean War – Communist China, in fact, had issued a grimly veiled warning: if United Nations crossed the 38th Parallel in force, she declared, then the Chinese people would not 'stand idly by'. Clearly the 38th Parallel had a dangerous importance in international affairs.

General MacArthur considered that Chinese intervention was unlikely so long as he did not advance right up to the Yalu, but halted some miles south of the Manchurian border. For the moment, however, the decision was out of his hands: aloof from the noisy arguments going on in the United Nations Assembly at Lake Success, MacArthur broadcast a demand for the surrender of all North Korean forces; this broadcast was backed by the dropping of 2,500,000 leaflets from United States aircraft with particulars of this demand.

This demand produced no reply from North Korea. A second appeal calling for immediate submission and cooperation with the United Nations, was also ignored. This was enough for

MacArthur: backed by the General Assembly's resolution of October 7th, he launched his forces over the 38th Parallel two days later.

[3]

It was not so much an advance as a headlong race: every unit of the American Army wanted to be the first to reach the Yalu, and they elbowed one another out of the way in this frenzied northward rush. As Eric Linklater wrote in his book *Our Men in Korea*, 'their impulsive hearts galloped far ahead of their roaring tanks and multitudinous transport'.

The men of 27 Brigade, who had arrived in Korea with no transport at all, never ceased to marvel at the amount of vehicles at the disposal of the American Army. On October 9th, as the Americans 'let 'em roll' in their frantic pursuit of the beaten North Koreans, they goggled afresh. For every road, amidst a cloud of choking, yellow dust was packed with tanks, trucks, lorries and jeeps nose-to-tail: there was no traffic control; vehicles double-banked and remained static, incapable of forward or backward movement, for hours on end; after the first frenzied excitement of the advance, progress of four miles in two hours became something exceptional. There were traffic jams which would have made the normal everyday snarl-up in London look like a hundred miles of clearway on U S Highway Number One. The progress of the Americans resembled a herd of buffalo on the rampage.

Miraculously jams were unravelled, and the vehicles roared and swirled ever onwards. There was no 'mopping-up' – huge dumps of North Korean arms and equipment were left far behind, as were bewildered enemy soldiers who found themselves by-passed by their conquerors: many of these surrendered (if they could find anyone static for long enough to accept their surrender), but many more, marvelling at their good fortune in escaping the advancing hordes, harassed road-bound convoys with vicious intensity and desperate courage.

During this frenzied forward move supply and maintenance services pleaded for a slow-up in the advance and pleaded in vain: there would, said the staff officers – themselves infected with the ecstasy of pursuit – be plenty of time for maintenance later. In the meantime, just keep going . . . keep going . . . keep going . . .

Brigadier Aubrey Coad rushed hither and thither in search of coherent orders for 27 Brigade's role in this advance, and finally discovered he was to take his brigade by air to Kimpo, near Seoul. On arrival there he would receive further orders.

The move of 27 Brigade from Taegu to Kimpo airfield was a model of its kind, for the entire airlift was handled by only two staff officers – one British, one American – and Brigadier Coad reports that not a single piece of paper was necessary. Men were assembled in batches of forty and shephered on to C47 aircraft with the minimum delay and fuss. Only one miscalculation was made: seventeen men were left standing bereft on the airfield, but within half an hour another aircraft had appeared. Aubrey Coad, ever sparing in his praise of American staff work, was constrained to describe the move of his brigade as a masterpiece of planning.

The deplaning at Kimpo proceeded equally smoothly. Argylls, Middlesex, and Australians were then driven to dispersal areas in Kaesong. Here 27 Brigade came under command of Major-General Hobart Gay's 1st Cavalry Division. Hobart Gay greeted Aubrey Coad like a long-lost brother: 'Always glad to have you boys with us,' he declared, striking a swift blow for Anglo-American friendship and immediately bringing a phalanx of newspapermen to the alert. He jabbed at a huge map with an impatient forefinger. 'Now, we've got a real nice assignment for you. Here's how it goes . . .'

Coad's assignment – it was an expression that he never really got used to, preferring to use the abbreviated and functional Anglo-Saxon word 'job' – was to take part in a forward movement to be carried out in conjunction with the 1st Cavalry and 24th Divisions. The role of 27 Brigade was to advance from Kaesong through the 7th United States Regimental Combat Team and lead the way up the Sariwon–Pyongyang Road; its objective was Sariwon, about thirty-four miles away, and reputed to be an important military training centre – the 'Aldershot' of North Korea.

Coad was delighted to be given a definite order at last. For him, so far, the 'race to the Yalu' had had little significance – 'a bloody shambles', was his own unequivocal description. 'Take Sariwon' was an order that a soldier could readily understand.

The advance to Sariwon was spearheaded by the 1st Argylls with Major David Wilson's 'A' Company in the van, travelling in lorries and sitting astride Sherman tanks.

The Argylls swept away the negligible opposition in the outskirts of the town with workmanlike precision: they were moving so fast that they hardly bothered to do more than machine-gun the few stout-hearted North Koreans who were manning road blocks.

Sariwon resembled a ghost town – a decaying mass of crumbling rubble, for it had been thoroughly bombed by the United States Air Force. To the bloated and rotting corpses littering the street, the Argylls had added another fifty. While the Argylls secured the town, the 3rd Battalion The Royal Australian Regiment, spoiling for a fight, pushed through the town to cut the main road from Pyongyang and the north. The Australians took up defensive positions to the north of Sariwon, and with volubly expressed disapproval and itching trigger fingers waited for the scrap which had hitherto been denied to them. They did not have to wait for long . . .

At first the scene in Sariwon was orderly enough, but towards nightfall the situation disintegrated into wildest confusion. The capture of the town had contained elements of high drama, but the later incidents smacked of low comedy, if not knockabout farce.

Some 2,000 North Koreans, retreating before the sledgehammer American advance, began to trickle into Sariwon, not knowing that 27 Brigade had occupied the town. The North Koreans, who badly needed friends at this stage of the war, encountered a party of Argylls under Lieutenant Robin Fairrie. The North Koreans eyed the Jocks warily: they were clearly not Americans, for they were not wearing helmets but some sort of woollen hats (the Argylls were wearing the knitted 'cap comforter', which was the standard headgear for British troops in Korea). Therefore, deduced the North Koreans, they must be Russians – their natural allies in the fight against capitalist aggression.

A North Korean officer approached Fairrie with outstretched hand. 'Russky?' he inquired.

Fairrie did some quick thinking: he was outnumbered by something like ten to one, and in a moment of lunatic decision he thought of singing *The Volga Boatman*. He settled, however, for the only Russian word he knew which was 'tovaritch', and said it loudly, clenching his fist in what he imagined was a Communist salute at the same time; at least, thought Fairrie uneasily, none of the Jocks of this particular platoon would admit to being English.

The North Korean extended the hand of friendship and the Jocks, entering into the spirit of the thing, made guttural noises in their throats. More handshakes and cigarettes were exchanged, and it seemed to Fairrie that they were well on their way to establishing an *entente cordiale* with the enemy, particularly when a young woman – 'and not a bad looker, either', recalled Fairrie – attached herself to him and exhibited every evidence of goodwill.

This uneasy situation continued for some minutes. Then Robin Fairrie, amazed at his luck but reluctant to push it too far, decided to ring down the curtain on a farce which could very easily turn into tragedy. It was time, he indicated with dramatic signs and still more guttural noises, to move on. It was then that a roving American in a jeep came round the corner and plunged both feet into it with sickening thoroughness.

The American was wearing a steel helmet, and the friendliness of the North Korean quickly disappeared. A North Korean officer pointed accusingly at the American and demanded: 'You Russky?' in a voice which sounded clearly disbelieving.

The American, who had been brought up to regard the Russians as his natural enemies, first opened his mouth in incredulous amazement and then went for his gun.

The ensuing scene was incredible, even by Korean War standards. The Argylls were ready for instant action – every automatic weapon was cocked – but the North Koreans, rendered careless by the meeting with 'Russians' were not. In the resultant flurry of wild shooting they fled in undignified panic, leaving a score of dead and wounded behind them.

It is impossible not to feel a certain sympathy for the luckless North Koreans who came to Sariwon, not knowing that the town was in Allied hands. They came by the truckload and on foot, from the south and the west: they were tired and dispirited, and wanted sanctuary, food, and sleep. They got none of these things.

As darkness fell, the situation, confused as it had been in daylight, became chaotic: North Koreans wandered about the town, not knowing where they were going or why, and were shot by prowling Argyll patrols who, after the 'Russky' débâcle, had been ordered to shoot first and ask questions afterwards.

Traffic jams were a daily feature of Korean roads during the October advance, but those in Sariwon were something else again: truckloads of 27 Brigade men passed truckloads of North Koreans, each unaware of the other's identity; on more than one occasion they were parked on opposite sides of the road. Major Ian Ferguson, the second-in-command of the Australian battalion, drove several miles in convoy, sandwiched between North Korean vehicles. Colonel Leslie Neilson, commanding the Argylls, drove along a road with his own 'B' Company on one side and North Koreans on the other, and came under heavy fire from both.

The North Koreans who managed to evade the Argylls blundered into positions to the north of the town held by the Australians, who slew them with grim enjoyment: some walked straight into machine-gun posts and were mown down; others, turning to flee, were brought down like rabbits by rifle fire – the 'Ratty Diggers', who had done this sort of thing before in New Guinea, found that they had lost none of their skill as marksmen.

All exits from Sariwon were most thoroughly sealed off; wherever the North Koreans went they encountered the same efficient slaughter.

The Australians, however, dismissed the Sariwon battle as a pretty tame affair – 'flamin' cut-sandwich commandos, that's us', said Private Jack Coultas, disgustedly, for the hapless North

Koreans were finally reduced to despairing wailing and surrendered in droves.

Eventually some semblance of order was restored to Sariwon. A few more North Koreans shouted 'Russky?' hopefully, but the Argylls were tired of this joke by this time and shot them; the Australians, who had not thought it particularly funny in the first place, dispatched a few more, including a particularly vociferous full colonel who was, as Lieutenent Eddie Larsen described him, 'loaded with flamin' red tabs and much good they did him'.

An exhausted silence descended on Sariwon. It had been a triumphant day for Coad's Brigade: they had advanced forty-two miles and, almost incredibly for the loss of one man, had killed 215 North Koreans and captured 1,200 more.

But there was to be no time for jubilation or relaxation; no time to sit down and count the score. For there was to be no stopping on this advance . . .

SEVEN

GENERAL DOUGLAS MACARTHUR had confidently predicted that the seaborne landings at Inchon would win the Korean War at a single stroke. Subsequent events had very nearly proved him right, but the unpalatable fact remained that in the third week of October 1950, the war was still very much on: in spite of a succession of resounding defeats (rendered even more so by the world's press), the North Koreans showed no signs of abandoning the struggle as the United States 1st Cavalry Division, pressing ever nearer to the North Korean capital Pyongyang, were finding to their cost.

American generals, however, were jubilant about the prosecution of the war, and gave the impression that it was all over bar the shouting – many of them, indeed, had started to shout already. General MacArthur prophesied that 'the war is very definitely coming to an end shortly'; General Walton Walker, deviating little from the shooting script, declared that 'the North

Koreans have been brought to a condition of collapse and disintegration from which there could be no recovery'; Major-General Hobart Gay, commander of the 1st Cavalry Division, had a telling punch-line to hand: 'The Communists,' he said, 'are all washed up.'

The wild optimism of the advance – 'cutback fever', as one American historian called it – had infected every American soldier in Korea; indeed, it did not seem that they would have to be there very much longer. Plans were afoot for the diversion of bulk-loaded ammunition ships to Japan, because it was thought that existing supplies in Korea would be sufficient; the 1st Cavalry Division was already formulating plans for a Thanksgiving Day Parade on the Plaza in Tokyo, complete with the yellow cavalry scarves which had been worn in bygone wars not so very materially different to this one; listing prices of goods available at the ever beneficent PX were light-heartedly thrown away – the soldiers were certain that they would do their Christmas shopping in Japan; MacArthur was already taking steps to establish a Civil Assistance Command to carry out the task of rehabilitating the people of Korea. Even the *New York Times*, which always made a special point of never making a statement which it might have to retract later, stated with editorial caution: 'Except for unexpected developments along the frontier of the peninsula, we can now be easy in our minds as to the military outcome.'

[2]

In order to ensure the final destruction of the North Korean Army, MacArthur planned a massive airborne operation on Sukchon and Sunchon: he estimated that something like 30,000 North Koreans would be trapped in a neat nutcracker between the airborne force and the United Nations advance northwards from Pyongyang. It was also hoped that a number of prominent North Korean government officials might be captured and that a trainload of American prisoners, in transit from Pyongyang, would be rescued.

For this operation – in its scope and magnitude it almost

equalled the Inchon invasion – the 187th Airborne Regimental Combat Team, commanded by Colonel Frank S. Bowen, had been held in G H Q reserve at Kimpo airfield. Bowen's men had been groomed for stardom and were flippantly known in some military circles as 'MacArthur's Own': if the Marines at Inchon had not completely put paid to the North Koreans, then the 187th Airborne certainly would in the area between Pyongyang and the two drop zones at Sukchon and Sunchon.

The paratroopers, each man a walking arsenal armed with .45 pistol, carbine or rifle, and a lethal-looking dagger, filed on to troop-carrying aircraft on the morning of October 20th. Their morale was high, for in common with every other soldier in Korea, they were infected with MacArthur's 'home for Christmas' forecast.

No chances were being taken with the 187th Regiment's drop: as the troop carriers approached the drop zones, fighter planes rocketed and strafed the ground with murderous thoroughness. Just after noon, Colonel Bowen and 1,470 men dropped heel to toe from 700 feet, south-east of Sukchon. During this and succeeding days approximately 4,000 men were dropped, together with 12 105-mm howitzers, 39 jeeps, 38 quarter-ton trailers, 4 anti-aircraft guns, 4 three-quarter-ton trucks, and 584 tons of ammunition, gasoline, water, rations, and other supplies. Against such an air armada, it seemed that no North Korean could possibly survive and General MacArthur, who had watched the drop from his personal plane, was jubilant. 'The airdrop,' he said, 'was an expert performance,' and this was undeniable. But his statement 'this closes the trap on the enemy' was scarcely supported by subsequent events.

The overall results of the airborne operation were, in fact, mortifyingly small: the 187th Regiment had not cut off any sizeable part of the North Korean forces – the main body of the enemy had already withdrawn north of Sukchon and Sunchon, and were either north of the Chongchon River or in the act of crossing it; no important North Korean government officials were captured, for they had all fled from Pyongyang eight days earlier; most of the American prisoners had been successfully removed into the remote fastness of North Korea – a search of the area revealed seventy-three dead American soldiers, who had either been shot or died of dysentery, starvation or exposure.

Although the 187th Airborne Regiment had, for the loss of less than a hundred battle casualties, killed over 1,000 North Koreans and captured 4,000 more, the North Korean Army was still in the war and soon made this abundantly apparent.

Meanwhile, 27 Brigade continued to lead the advance to the north: they crossed the Taedong River at Pyongyang and headed towards Sukchon. Approaching Yongyu sounds of furious battle only a few miles away could be heard – clearly the North Koreans were keeping the airborne men busy. The light was fading and the men were tired; like a man invited for cocktails at six who finds that he can't get there before seven-thirty, it seemed to Aubrey Coad that this battle was a party which he was not called upon to join in, at any rate not in pitch darkness and with no idea where Americans or North Koreans were. To the audible disgust of the Australians who, like Job, smelled the battle afar off and wanted to be in it, Coad halted the brigade for the night.

The sounds of battle continued throughout that night, making sleep for the men of 27 Brigade virtually impossible. Most of the noise came from the 3rd Battalion of the 187th Airborne Regiment who were engaged in the efficient slaughter of the North Korean 239th Regiment, which had been left behind in the general enemy withdrawal towards the Chongchon River: the airborne men killed 805 North Koreans and captured 681.

At first light on the 22nd the 1st Argylls advanced cautiously into Yongyu in the face of desultory sniper fire. They dealt swiftly and peremptorily with the snipers; then, looking for fresh targets but finding none, the Jocks set about burning Yongyu to the ground with obvious enjoyment.

Hard on the heels of the Argylls came the 3rd Battalion The Royal Australian Regiment, riding on American tanks. In the words of Sergeant Bill Coffman, 'they were at concert pitch and screaming to go'.

The Australians were belligerent and frustrated, and a

belligerent and frustrated Aussie invariably spells trouble for someone whether on a battlefield or in a saloon bar. They had come to Korea to fight, and apart from Sariwon – primarily an Argyll battle and dismissed as a 'pretty tame do' by the Australians – they had done all too little so far. They had expected to be committed to a full-scale attack, but the spectacle of 2,000 North Koreans tamely surrendering had seemed to sound the knell of battle. Rumours of Australian dissatisfaction had percolated into 27 Brigade Headquarters, and Brigadier Coad fervently hoped that they would run into a fight before long. Fortunately for his (and Colonel Charles Green's) peace of mind they did so that very afternoon.

Just north of Yongyu, where the Australians had exchanged ribald pleasantries with the Argylls, the leading tanks carrying Captain Pat Denness's 'C' Company ran into heavy rifle and light automatic fire from an orchard that spread out on both sides of the road; the North Koreans were well dug in, and the Australians on the tanks presented tempting targets. Without waiting for orders from anyone or pausing to appraise the enemy strength, Denness led 'C' Company straight into the attack with bayonets fixed.

'C' Company were soon under fire from three sides, and Colonel Green, realizing that 'C' Company couldn't do it on their own (although afterwards they claimed that they could have), quickly deployed the other three companies. Soon the entire battalion, like currants spilling from a basin, were swarming up the hills on both sides of the road; this was something like the situation for which the Aussies had been waiting.

Coad and Green both knew that the American airborne troops were less than a mile away, but they had no method of pinpointing their exact positions. Consequently, no artillery or mortar fire could be brought down in support of the Australians for fear of hitting the Americans. The Australians were not quarrelling with this order: in this and subsequent battles they gave the impression that they did not care if they had artillery support or not.

The fighting had to be done with infantry weapons and nothing else: the bren gun, the hand grenade, the rifle, and the bayonet – particularly the bayonet, a form of fighting in which the

Australians have always excelled. Blasting and bayoneting the North Koreans from their trenches; fighting with clubbed rifles, boots, and fists; hacking, stabbing, and cursing, the Australians charged ever forward. Some of the braver North Koreans stayed in their trenches and fought it out where they stood until they went asunder. Others, less stout-hearted, fled before the bayonets of the Australians into range of the American tankmen, who slew them with relish. Undoubtedly a number of North Koreans attempted to surrender, but the Australians had seen enough prisoners to last them a long, long time and none were taken in this operation.

Enjoying himself hugely throughout the battle – indeed, he gave the impression that it had been laid on for his express entertainment – was Private 'Bluey' Smith of 'C' Company.

'Bluey' was a 'Ratty Digger' *par excellence*. He had joined the Australian Army in the early days of the Second World War, and after one year of fractious home service his conduct sheet had come to resemble something like a full-length novel – in the absence of a tangible enemy, he had waged a private tribal war with NCOs and military policemen. It was not until he reached New Guinea that he discovered his natural *métier*, as a score of notches on his rifle butt, each representing a dead Jap, bore grim witness.

He stood six feet and two inches in his bare feet and weighed two hundred pounds, of which possibly two were superfluous. He had a face which looked as if it had been through a meat grinder, a voice like a truck in low gear, and flaming red hair (in Australia red-haired men are nicknamed 'Bluey'). Here was a natural fighting man, but to the North Koreans and several of his comrades it seemed that 'Bluey' was the devil himself.

He was certainly no bad substitute. From the moment that he jumped from one of the leading American tanks, 'Bluey' was a one-man wave of destruction. Very early in the attack, roaring like a bull and twisting his rifle round his head like a cane, he fell on a party of North Koreans crouching in a trench: he shot three men at a range of about three feet; he disembowelled two more by thrusting with his bayonet and ripping upwards; the other men in his section heard rather than saw the blow that fetched home on the skull of a North Korean who may or may not have been contemplating surrender. There had been eight enemy soldiers in this particular trench, but this wild apparition which was 'Bluey' finished them.

Having satisfied himself that the battle was going well, Charles Green moved his battalion headquarters forward. They were immediately attacked by a party of some sixty North Koreans, who had somehow been by-passed in the Australian attack. Standing back to back, Green, his signals officer, Captain Laurie Watts, his batman, Private Jack Leyman, and a dozen signallers killed thirty-four of them for the loss of three wounded. Afterwards Private Leyman said ruefully he had volunteered for the job of C O's batman because it was a 'soft number'.

In spite of having been virtually routed, the North Koreans were not giving in easily. Those who survived the Australian attack fled from the hills to the west and hid in paddy fields, ditches and rice stooks. From these vantage points they kept up a vicious and petulant sniping fire on the Australians as they consolidated the ground which they had so bloodily won.

However, these snipers were not a nuisance for long. A party of Australians about thirty strong, with the considerable addition of Private 'Bluey' Smith, who had absent-mindedly wandered into the wrong platoon, lined up in a paddy field and walked straight across it – 'for all the world,' said Coad, who lists shooting among his hobbies, 'as if they were driving snipe. The soldiers when they saw a pile of straw or a rice stook, kicked it and out bolted a North Korean. Up with a rifle, down with a North Korean and the Australians thoroughly enjoyed it. They did that the whole day and were absolutely in their element.'

The 3rd Battalion The Royal Australian Regiment's first really tough action had been crowned with success, and brought forth unstinted admiration from everyone who witnessed it. In this hand-to-hand snarling dog-fight 270 North Koreans were killed; incredibly, the Australians only suffered seven wounded – the original figure had been eight, because a medical orderly had seen

'Bluey' Smith covered with blood and with his clothes slashed to ribbons. 'Wounded, am I?' said 'Bluey'; 'you ought to take a look at the other jokers.'

As well as being the most reported war in history, Korea was also the most photographed. Cameras were largely an American import to the battlefield: in addition to combat cameramen, who faithfully and courageously recorded the war on celluloid from Pusan to the Manchurian border, it seemed that approximately one American soldier in two owned a camera, and many a British soldier acquired a handsome Leica in exchange for a carefully hoarded liquor ration; several Australians, photographic ambition struggling with national pride, parted with their slouch hats, which had a particular fascination for American soldiers.

The original 27 Brigade arrived in Korea with very little of anything, and certainly no cameras – not for nothing were they known as the 'Woolworth Brigade'. But the 3rd Battalion The Royal Australian Regiment, in addition to no fewer than six chaplains (including a Salvationist), boasted a unit photographer in their strength. Wherever the 3rd Royal Australian Regiment fought, Private Ian Robertson was on hand with his camera to record the occasion – sighting the subject, dodging bullets and the invective of officers who had other things to do besides posing for pictures, securing photographs of exemplary perfection.

Robertson had been a cameraman on one of the Sydney dailies when his attention was attracted to a recruiting poster. He soon became very popular in the battalion, for Robertson knew that soldiers, whatever their rank and however much they might deny the fact, love having their photographs taken. The constantly repeated shout, 'Hey, Robbie, how about a snap?' resulted in the expenditure of hundreds of feet of unauthorized film – usually with the subject looking suitably belligerent with bayonet fixed, and whenever possible standing over a dead North Korean.

On October 21st, however, Robertson was called upon to show his prowess as a soldier as well. During the attack a North Korean, feigning death, suddenly came to life and hurled a grenade and the unit cameraman took a splinter in the wrist. Moved to furious rage, Robertson temporarily substituted rifle for camera and shot the North Korean dead; then, as an afterthought, he photographed the corpse. Back in Australia, he unsuccessfully attempted to convince his colleagues on the paper that it had been all his own work . . .

The North Koreans were shortly to get another salutary dose of Australian medicine. A brigade commander with three battalions at his disposal must never heap too much dirty work on any one battalion; no one knew this better than Aubrey Coad, but he had observed that the Argylls and Middlesex were luckier than the 3rd Battalion The Royal Australian Regiment, because when he put the 3rd R A R in the lead they inevitably met tough opposition: during the month of October it seemed that the Australians attracted trouble in a way which was almost magnetic, and they would not have had it otherwise – this, after all, was what they had come to Korea for.

The next task of 27 Brigade was to secure a bridgehead across the Chongchon River at Sinanju and to capture Pakchon and Chongju, two towns on the main routes from Pyongyang to the Manchurian border.

In this particular advance the 1st Middlesex were in the lead, and crossed the Chongchon without interference from the enemy, but plenty from the elements and the river itself. The bridge over the Chongchon River just beyond Sinanju was down, and the Battalion was ordered to lead the Brigade in an assault of the river. In order to gain surprise, it was decided to make the crossing four miles downstream, where the river narrowed slightly. At this point it was about 800 yards wide, with an approach over open paddy of a further 1,000 yards and with low hills on the far side. The Middlesex had a rough trip on this river crossing: the first two companies got across without mishap, but the last of the assault boats in which the third company crossed grounded in mid-river and some fifty men had a nightmarish struggle to the opposite bank through thick and gelatinous black mud. The fourth company fared even worse, for by this time the tide was racing out and the Chongchon became a mass of small mud islands. The Middlesex spent the night of October 24th in horrible discomfort, exposed to everything that the increasingly foul weather had to offer; they also spent it without food or covering, for the rations and blankets were on the far bank of the river.

The crossing of the Argylls at Pakchon presented few problems, although a piper fell into the water and all but surrendered his precious bagpipes to the treacherous river. The Australians passed through the Argylls in Pakchon and then, in a manner which was to become almost traditional in 27 Brigade, ran slap into a furious battle.

Major Bill Brown's 'D' Company stormed into one attack, while Captain Bill Chitts took 'A' into another: both assaults were made in the face of strong enemy resistance, which was brushed aside in the lethal manner which was to make the fame of the 3rd Royal Australian Regiment secure in Korea for all time. Pat Denness's 'C' Company (and Private 'Bluey' Smith) were in reserve, but 'D' found themselves in the path of a frenzied counter-attack, in the 'banzai' style, launched in a desperate attempt by the North Koreans to dislodge the Australians.

Lieutenant Dave Mannett, commanding one of 'D' Company's platoons had detected a tendency among some of the old hands to fire wildly at nothing at extreme range: the fire discipline, sternly inculcated in Australian soldiers in the Second World War, had become exposed to rust. 'If any of you jokers go blazing off at flaming spooks in the darkness or fire before I give the order,' Mannett told his platoon, 'I'll have his guts for garters.' Thirty dead North Koreans, all killed at a range of approximately ten yards in front of Mannett's position, were proof that his words had taken effect.

A similar North Korean attack, spearheaded by two Russian T-34 tanks, met no better fortune against Captain Chitt's 'A' Company: Private Len Simpson, who unlike 'Grouser' Peck of the Middlesex, actually enjoyed humping a bazooka about Korea, destroyed two of the tanks with unerringly-aimed rockets; Private Jack Stafford disposed of yet another with a well-placed burst of bren-gun fire in its auxiliary petrol tanks – the resultant uproar of cheers was of the type which is heard when an English wicket falls in a Test Match at Sydney.

At daylight on October 24th the Australians were securely in their positions, and the gleeful task of counting corpses had started afresh – another fifty were added to the already impressive bag of the 3rd Royal Australian Regiment, even without the assistance of 'Bluey' Smith.

By one of those freakish twists of fate, inescapable in war, the 3rd Royal Australian Regiment lost their Commanding Officer, Lieutenant-Colonel Charles Green, at the very moment of victory: the last North Korean pockets of resistance had been overcome; the last shots had been fired at the fleeing enemy; the last few hangdog prisoners had been shepherded away from the battlefield.

Charles Green had earned a rest if anyone had. Throughout the action he had been undisputed master of the battle: calm, unruffled, almost irritatingly casual under fire. In the desperate close-quarter fighting when his battalion headquarters had been attacked, he had shown himself to be a deadly marksman. Now the battle was won, and his men could relax a little; they had been continuously on the move or fighting since they crossed the 38th Parallel twenty-two days earlier.

Green first toured his company positions: there were many things for a C O to do before he could seek that blissful aftermath of battle, a long and undisturbed night's sleep: he had to make sure that all companies were prepared for counter-attack, although at present this seemed unlikely; the morrow would bring all the administrative chores – casualty and ammunition returns, recommendations for awards, letters to relatives of the fallen.

No one knew where the solitary shell came from. There were something like a hundred Australians on the hillside – cleaning weapons, burying North Korean dead, brewing tea, smoking, and exchanging battle reminiscences. Colonel Green, on the urgent advice of Jack Leyman, his batman – 'Get your head down, Skipper,' had been the executive order – had gone to his tent and was asleep almost before his head touched the rolled-up greatcoat which did duty as a pillow. Then there was a single, screaming whine as the shell came down, followed by the explosion as Corporal Vincent Thompson and Private Kingsley Jones, who were a bare twenty yards from the Colonel's tent, flung themselves flat. Mercifully, it seemed that no one had been hit.

But a single splinter sped through the opening in Charles Green's tent and embedded itself in his stomach. He died that night, bringing the total Australian casualties to nine killed and thirty wounded.

It was a tragic ending to a victorious day; it had been a splendidly fought battle in the true traditions of Australian dash and courage – a fight to rank with actions fought at Gallipoli, on the Somme, in North Africa, and New Guinea. Charles Green had seen much death in New Guinea, and had exposed himself fearlessly to it there and in Korea. When death came to him at Chongju, it found him well content and secure in the knowledge that he had put Australia squarely on the map in Korea. His men mourned him bitterly and briefly, and then, like the tough soldiers that they were, put him out of mind.

[3]

At last the men of 27 Brigade were getting the rest for which their tired and aching bodies craved. Just twelve miles short of the Yalu River, they took off the remains of their cracked boots, cut away their foul and stinking socks and waggled their toes ecstatically, amazed to find that they had five toes on each foot. There were hot baths in 44-gallon petrol drums, and sergeant-majors urged unit barbers into action. Vehicles, which threatened to fall to pieces if driven another mile, were repaired. A liquor ration appeared, which was consumed in record time, and the beneficent P X showered 'goodies' on them with joyous abandon.

Letters were written home, and Colonels Neilson, Man, and Fergusson (who had succeeded Charles Green in command of the 3rd Royal Australian Regiment) wrote belated citations for gallantry awards.

On the night of October 30th there was almost a holiday spirit in the air. The Middlesex sat round fires and sang *Nellie Dean*, while the Australians countered with a raucous rendering of *Waltzing Matilda*. Somewhere in the Argylls' lines a lone piper played *Hey, Johnnie Cope, are ye wakin' yet?*

It was a night of strange, eerie beauty – the moon was up and the light was silvery – as Colonel Richard Stephen's 21st United States Infantry Regiment, an outfit of enduring toughness which had been in Korea from the very start of the war, went through 27 Brigade to resume the pursuit to the Yalu. Morale in the 21st was high: they could just hear the rumble of the North Korean tanks in retreat, and Thanksgiving Day in Japan seemed a virtual certainty.

And for the 21st, too, it was success all the way. Five hundred North Koreans, in a final desperate gesture of defiance, attempted an ambush on Lieutenant-Colonel Gines Perez's 2nd Battalion. By dawn on the 31st the North Koreans had fled in disorder, leaving behind fifty dead and seven knocked-out tanks.

The North Koreans fared no better against Lieutenent-Colonel Charles B. Smith's 1st Battalion, who killed one hundred more. This was the northernmost action fought by a unit of the United States 8th Army in the Korean War, and to the men of the 21st it seemed that it must be the last . . .

To the men of 27 Brigade it seemed a good opportunity to think of home, to sentimentalize and mourn a little, to reminisce, to mull over past mistakes and triumphs: to remember the Naktong crossing, Plum Pudding Hill, Sariwon, Hill 282, Yongyu.

It also seemed a good time to have some second, and in some cases, first thoughts about those brilliant, courageous, volatile, and unpredictable fighting men – the American soldiers.

[4]

Like a hopelessly incompatible married couple, who only have sex to sustain an uneasy union but are determined to make a go of it somehow if only for the sake of the children (in this context the future of the democratic free-thinking world), the British and American, strove towards a better understanding of one another.

It was inevitable that to the British every American soldier was a 'Yank', whether he was a steelworker from Pittsburgh or a corn-bred, buttermilk farm boy from Alabama. By the same reasoning, 'Limey' was the generic epithet for a British soldier from Penzance or Pittenweem.

The two national conceptions of discipline were widely different: the British soldier, seeing the cultivated American slouch and listening to the casual acceptance of orders, said 'Blimey, wish I was in a cushy mob like that'; the American, observing the salutes and hearing the crash of ironclad heels, remarked without envy: 'Hell, this is a rugged outfit.'

These differences, superficial to an unthinking British private (approximately one in two), were symptomatic of something more basic and are grounded in history: in the United States Army, with the exception of the West Point graduates, there was no officer caste *per se*: born of a bewildering mixture of races and bred in the backwash of colonial independence, American soldiers, from generals to buck privates, were so alike in speech and habits that only emblems of rank distinguished them.

The British soldier in Korea, as in every other war, spent a fantastic number of hours digging. G Is looked with pitying astonishment at British soldiers carrying picks, shovels, and entrenching tools; in their army, they explained indulgently, such impedimenta went on the truck. And it was no earthly good exclaiming, as did one short-tempered private of the Middlesex: 'We ain't got no f— trucks, mate, that's why I'm humping this b— lot.'

It seemed to the Argylls and Middlesex in the early days that, in their reluctance to dig and general unpreparedness, the Americans were asking to get killed.

Second-Lieutenent Stanley Fothergill of the 1st Middlesex was detailed to take over a position from an American company. The enemy were less than two miles away, yet only a single American sentry, a bored-looking individual smoking a cigarette, taking long refreshing draughts from a bottle of Coca-Cola and reading a comic, stood (or rather sat) watch and ward over the outpost.

Fothergill checked his map to make sure that he had in fact arrived on the right hill, and then addressed the sentry. 'Is this Hill 116?' he inquired.

'Search me, mac, I'm a stranger here myself,' said the sentry – he jerked his thumb towards a huge marquee, of the type that one sees at flower shows – 'I'll go ask the lootenant.'

'Where is he?' inquired Fothergill.

'In his sleeping-bag along with the other guys.'

The American lieutenant, unwillingly roused from sleep, was scarcely more forthcoming. With a slight edge of sarcasm, Fothergill asked him if he had ever thought of digging any trenches, for he had been under the impression that he was to take over a *defensive* position.

'*Dig*?' said the lieutenant, as if he had heard a dirty story in mixed company; 'what in hell's the use of digging if we're gonna have to bug-out?'

Stanley Fothergill, mindful of his orders to hold the position to the last man and the last round, let it go at that – and got his men digging.

Lieutenant Sandy Boswell of the Argylls encountered a similar situation, with one essential and macabre difference: the Americans he had been sent to relieve were dead to a man, and securely zipped into their sleeping bags.

Neither General John Church of the 24th nor Hobart Gay of the 1st Cavalry Divisions made any secret of their admiration for the officers and men of 27 Brigade and communicated this admiration to Brigadier Coad unreservedly. They naïvely confided to Coad that the presence of British troops was a steadying influence on 'kids who'd never seen combat before'. 'This,' said Coad wryly, 'was gratifying, but repetition can become tedious. It probably explains why we were nearly always caught for advance guard or rearguard.' In any case, the majority of the Argylls and Middlesex were 'seeing combat' for the first time, although the Australian battalion had a high proportion of Second World War veterans in its ranks.

There was one American unit, the 2nd Chemical Mortar Battalion commanded by Lieutenant-Colonel Edgar Bell, which formed the closest and most abiding friendship of all with 27 Brigade. Of Coad's men Bell had this to say: 'We were attached to the 27th British Commonwealth Brigade and I should stand up and bow. In all my experience I have never enjoyed an attachment so professionally perfect; they are the finest troops in the world and the finest men.'

Perhaps the closest affinity of all was between 'Charlie' Company and the 1st Argylls. And this warm friendship was initially engendered by the simple act of shaving in the morning.

Meet Captain Charles Snow of New Jersey; he had landed on Omaha Beach on D-Day and was in there pitching at the finish; he weighed two hundred pounds, give or take a four-course dinner; his three primary passions in life were his mortars, Scotch whisky, and poker playing. Together with his four forward-observation officers, Lieutenants Tom Holcomb, Tom Kolbes, Dale Henry, and Bill King, he presented himself to Major John Sloane, second-in-command of the Argylls.

Sloane expressed himself delighted to see them, as indeed he was, but mindful of his Commanding Officer and Brigadier he decided that in their present form the officers of 'Charlie' Company would not do at all. They saluted dramatically; they looked cheerful and professionally competent; they represented something that Colonel Neilson had been praying for – a heavy support fire unit. But each of their faces bore the thick black stubble of upwards of five days.

Brigadier Coad had given the order that the men of 27 Brigade would present a clean chin to allies and enemy alike. Therefore, Sloane said to Captain Snow: 'I think you should know that we shave every morning in this battalion,' and five American mouths dropped open in comical dismay.

Back in his bivouac, Charlie Snow examined his face in a few inches of cracked mirror and decided that he looked a disgusting object; furthermore, the stubble of four days was beginning to itch. He decided that he did not look tough – many soldiers think that several days' growth on the face imparts an aura of ruggedness – but merely dirty and unsoldierlike. He summoned his orderly, who was also generously stubbled, and commanded: 'Go get me some hot water.'

'Sure thing, Captain,' said the orderly. 'You fancy some cawfee?'

'I do not,' said Snow. 'I'm gonna have me a shave – and that goes for you too . . .'

From that day every man in the Mortar Battalion shaved every morning.

They did not only shave; they also took to drinking tea instead of coffee: Lieutenant Albert C. Huggard installed a billycan hanging from the towing hook of his jeep, and regarded it as the most essential item of his equipment. Those men who could, acquired British regimental badges and wore them proudly on their chests.

There were many other things which the British grew to admire in the Americans: apart from the 2nd Chemical Mortar Battalion, whose shooting was always a model of slick precision, 27 Brigade (who had no artillery of their own) acknowledged that in matters of gunnery the British could learn something from the Americans; aside from the tragic incident on Hill 282, the air support was magnificent and the American pilots were brave to the point of being suicidal; the tanks, too, answered all appeals for help and roared into the attack with unquenchable *élan*; the helicopter pilots swooped down to pick up wounded under murderous fire and won the admiration of all who saw them; the Engineers, often under heavy fire, threw pontoon bridges across rivers with carefree abandon and seemingly miraculous efficiency and speed; the quartermastering was always on a lavish, if not profligate, scale.

Perhaps the best summing-up of growing mutual liking and respect came from Captain Charlie Snow, while he was sharing a fortuitous bottle of Johnny Walker with Major Sloane: 'Hell, Johnny,' he said, 'we don't do things better, we just do 'em different.'

EIGHT

THE month of October 1950 had produced success for the United Nations all the way, and the Americans looked forward to their Thanksgiving Day, November 23rd – spent, if not in the United States, at least in an atmosphere of victory. From early colonial times Thanksgiving Day was celebrated in recognition of such happy events as good harvests and victories over the Indians, it is a day traditionally given over to lighthearted rejoicing to the accompaniment of gigantic consumption of turkey, and if the harvest in Korea had been meagre the victories over the North Koreans had been sensational. There was ample cause for celebration, both in the religious and the secular sense.

But November produced little cause for thanksgiving; it produced instead almost unrelieved disaster. Retreat, defeat, vile climatic conditions, misery, slaughter, famine, pestilence – the Four Horsemen of the Apocalypse appeared to have been increased to a full-strength squadron of cavalry.

Initially Brigadier Aubrey Coad felt no undue sense of foreboding as he set out for Divisional Headquarters of the 1st United States Cavalry Division on the evening of November 2nd. He was, he freely admits, looking forward to one thing only – a period of extended rest for his brigade, and incidentally for himself.

Divisional Headquarters was more crowded than usual – the smoke thicker, the babel of conversation on a more high-pitched note; everyone seemed to be talking at once, but no one appeared to be listening.

Coad pushed his way through groups of vociferous staff officers, orderlies scurrying to and fro with signals and fresh supplies of coffee, and newspaper correspondents – the press contingent alone seemed to have doubled in size since the previous day. Eventually he found General Hobart Gay, standing in front of a huge map. One look at Gay and the map told Coad that all was very far from being well: the map looked as if it had been assaulted by a lunatic artist armed with a red chinagraph pencil; from every direction arrows of brilliant red defaced its surface.

'The Chinese are in,' said the general to Coad, without preamble, 'the Third World War has started!' Aubrey Coad, wryly humorous and infinitely quotable as always said afterwards: 'I thought that a good beginning for any evening.'

The hammer blow had fallen. The war, which seemed virtually finished, was only just starting.

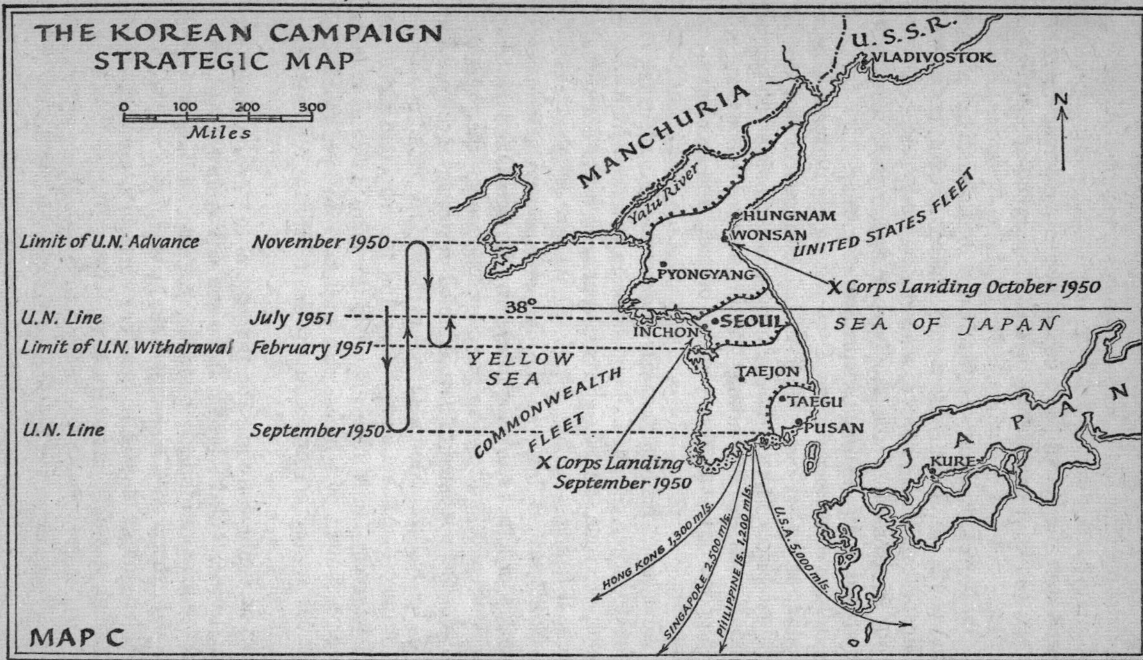

©CASSELL & CO LTD 1967

[2]

Simultaneously with the Chinese intervention came the cold – cold such as the British Army had not been called upon to fight in since the Crimea, a century earlier, and the amenities in Korea were not appreciably better; there was no Florence Nightingale in Korea, or anyone like her – if the ladies of Korea carried any lamps, they were certainly not for the soldiers.

On the night of November 14th the wind came howling in from Siberia by the most direct possible route, and the temperature dropped to 22 degrees of frost (on the coldest night on record the thermometer registered 45 degrees). The whole of Korea was enveloped in a deep freeze: where before the seeping yellow dust had been the greatest curse, now drizzling rain settled the dust and then froze into a hard glaze over the rough mountain tracks; the biting wind moaned down the frozen valleys and glanced like icy daggers off the frost-sheeted paddy fields; rivers became glittering slabs of ice.

It froze water and every other form of liquid: water poured from a jug froze before it reached the ground; boiling water poured into a vehicle's radiator became a block of ice almost at once. The mention of anti-freeze brought a hollow laugh from transport drivers; there *was* anti-freeze, an unclassifiable substance of Russian manufacture which had been captured from the North Koreans and was allegedly infallible in the severest winters. It did nothing to ease the always parlous condition of 27 Brigade's transport – not to put too fine a point on it, it froze.

It froze tea, beer, and eggs in their shells: a man who absent-mindedly put a mug of tea down for ten seconds found that it had turned to brown ice; Americans and Australians favour ice-cold beer, but this was carrying things too far altogether; an egg was a rare luxury in Korea, but a yolk frozen into a hard yellow ball was scarcely calculated to coax the appetite. Cooking, never an enviable chore in Korea, became an ordeal of the most daunting description: on British pattern cookers water took 1½ hours to boil, and a tin of meat to become even lukewarm had to be immersed in boiling water for at least two hours.

It froze weapons, tanks, and razor blades: oil, which kept rifles and machine guns in serviceable condition, froze at the first application; bare fingers froze to the sides of tanks and to razor blades – shaving became a torment, and usually large quantities of skin were removed with the beard; men with moustaches shaved them off when they found inch-long icicles adhering to the hair.

It froze men: sentries had to be relieved after half an hour's duty – any further exposure to the elements could (and did) result in men being found frozen as stiff as a plank of wood; to combat frostbite, socks had to be changed twice a day, but they could only be put on and taken off if the wearer was near a fire. Breath froze; men froze to their rifles; soldiers after riding in a jeep fell flat on their faces when trying to walk; blood literally froze – inside and outside the body; men who had been wounded were found to have a solid red block at the point of entry of bullet, shrapnel, or bayonet. Dark objects sticking out of the snow were found to be the toes of dead men – the quick freezing had turned them claret-red.

To keep alive, the soldiers had to learn the most primitive tactics of survival. Everywhere men were frantically fossicking for something to burn: they tore up railway sleepers, hacked down telegraph poles and gathered rice sheaves with which to line their trenches.

The Australians were hardest hit of all, because prior to coming to Korea few of them had ever encountered snow before; at the beginning of the cold spell they indulged in a riot of snowballing, but not for long; soon the cold pierced deep into the marrow, causing every joint to ache unbearably.

There were no malingerers in the 3rd Battalion The Royal Australian Regiment, quite the reverse: many lightly wounded men scorned medical attention and remained at duty. But the murderous cold was seeking out the weak spots among the older men, many of whom had cheated on their ages when they volunteered. The battalion medical officer, Captain Ben Gandeira, was kept busy with men doubled up with rheumatism, backaches, lumbago, and kidney trouble.

Remarked a shivering Australian soldier: 'This is no country for a white man.'

'It's no country for a black man either,' countered Lieutenant Reg Saunders, the only Aborigine ever to be commissioned into the Australian Army.

Mercifully, warm American clothing put in an appearance, and for the Argylls and Middlesex it came just in time to save them from literally freezing to death; when the cold first struck they had been wearing the same tropical uniform in which they had left Hong Kong.

As in all aspects of equipment issue, the Americans did a fine job: there were string vests and woollen 'long-johns'; extra heavy pullovers; oiled socks and frostproof ski-type boots; loose-fitting wind-proof suits and fur-lined ski-caps with ear flaps. But no amount of Eskimo furs or Manchurian quilting could have kept the men weatherproof if their spirit had failed: the British soldier is the most adaptable creature in the world, as he has proved in deserts, jungles, and swamps; in Arctic conditions he was not found wanting.

But the overall portents, depressing as they seemed, were not entirely black, for men were coming to Korea from almost every free country in the world to aid the United States, Great Britain, South Korea, and Australia .Indeed, men of so many nationalities appeared that it seemed that this was in truth becoming a world war.

Came a New Zealand regiment of artillery, a cheerfully amateurish bunch: apart from the senior ranks, hardly any of them had handled an artillery piece prior to the formation of the regiment four months earlier. Many of them had seen service in the Second World War as almost everything except gunners – there were ex-sailors, ex-fighter and bomber pilots, ex-infantry-men, ex-tankmen.

But there was no place for amateurs in Korea, and the New Zealanders soon showed that in a bare four months they had achieved a remarkable degree of competence.

There also arrived a redoubtable Canadian battalion, with a fine fighting record, the 2nd Battalion Princess Patricia's Light Infantry: at Pusan they were greeted by the now permanent Negro band playing *If I'd known you were coming I'd have baked a cake*. As far back as August 7th, 1950, the Canadian

Government had informed the United Nations that it was prepared to place a special force at the disposal of the United Nations Command in Korea. The 'Princess Pat's' were the forerunners of many other Canadians who were shortly to lend their weight to the Korean War.

Like the Australians, the 'Princess Pat's' were spoiling for a fight; like Charles Green, their Commanding Officer (Lieutenant-Colonel John Stone, DSO, MC) intimated to Aubrey Coad that it couldn't come quick enough. Like the Australians, they did not have to wait long

The 60th Indian Field Ambulance arrived, and soon earned pride of place as the finest medical unit in Korea – an assessment with which no soldier of any nationality ever quarrelled. Commanded by Lieutenant-Colonel Rangaraj, a doctor first and a soldier second and immune to internal Indian political strife, every man was a veteran of the Burma campaign, a qualified parachutist, and a dedicated healer. Soon British, American, and Australian sick and wounded were openly declaring that they wanted the Indians to minister to them, and many a British soldier surfeited with canned hash, sampled a curry of Veeraswamy potency and proportions. There were no 'ladies with lamps' in Korea, but the 60th Field Ambulance were the next best thing.

There were soldiers in Korea with white, brown, and near-black skins; soon there were man with olive-coloured skins too – there came a thousand Greeks of the same calibre as the men who had caused Mussolini some second thoughts about Fascist superiority in Albania: their officers had been specially selected for their knowledge of English, and every man was a highly trained and rugged anti-Communist fighter, and no stranger to tough mountain campaigning in the post-war Communist troubles in their homeland.

Came a Norwegian surgical team – serious, blond men who were acclimatized to foul winters, and knew all about a country tyrannized by a vicious invader. There was a French infantry battalion, and they first earned the exasperation but later the admiration of their American hosts: the Frenchman, in any circumstances, appreciates warmth, and in the grisly Korean winter the men of this battalion tended to light over-lavish fires. A distracted American staff officer, observing a hilltop position bathed in comfortable flame, issued a vitriolic instruction to the Commanding Officer: 'For the love of God, get those goddam fires out or you'll have all the Chinese in Korea round you.' '*Magnifique*, *monsieur*,' was the totally disarming *riposte*; 'as you say, the Chinese will see our fires; they will attack us and we will kill them.' And so they did . . .

There was a Belgian battalion, imbued with the spirit of Burgomaster Max of Brussels in 1914, 'who faced the German bullies with the stiffest of stiff backs', and were to meet the Chinese in the very same way. Small, grinning Filipinos, of the type who had been a thorn in the side of the Japanese after the fall of Bataan, arrived thirsting for a fight. Colombians, Bolivians, Ethiopians, Dutchmen, and Luxembourgers were equally determined that Communism would never gain the whip-hand in Asia. South Africa, not famed for love of Britain or America, sent a fighter squadron and a team of army staff officers.

The arrival of the 21st Thai Infantry Regiment at first produced a situation of rich comedy. A tubby-looking little man, of undeniably Oriental appearance, reported at the Headquarters of an American Regimental Combat Team for orders. The Combat Team's executive officer, a cynical soldier who by this time was certain that he'd seen it all, announced laconically to the Commanding Officer: 'There's a gook outside, Colonel.' The Oriental officer, who had heard this impeachment but was not prepared to admit it, stepped briskly into the command post. 'I am not gook,' he corrected with a charming smile; 'I am commander of Thai soldiers. Where are the gooks we have come to fight?'

Last, but by no means least, came possibly the toughest soldiers of them all. Ask a vintage British old sweat who are the most formidable enemies he has ever fought, and you will get a variety of answers: two generations of British soldiers will unreservedly name 'Jerry'; few will argue with 14th Army men who firmly say 'the Nips', and stifle any further discussion on the subject. But the old warriors who fought at Gallipoli and in Mesopotamia will say 'Johnny Turk', or 'Abdul', as he was often known, and they are probably right: hard, frugal, and cruel – murderous trench fighters and snipers; only death could stop them in attack; in defence they held their ground with terrifying tenacity.

And now the Turks had come to Korea and this time to fight on the right side.

Their commander, General Tashin Yazici, had fought against Allenby in Palestine; Colonel Sirri Acar, as a 16-year-old platoon commander at Gallipoli, had killed Anzacs in 1915, as had the fathers of most of the men under his command. This in no way affected an instant *rapport* between Australians and Turks when they first met in a miserable little mud-and-stick village called Kunu-ri in Korea: the seeds of mutual respect had been planted at Gallipoli, and they ripened in the snow of Korea; Turks and Australians, numbed with the cold, squatted round fires – the Turks experimented gingerly with American chewing gum; the Australians chewed manfully on something that looked like pancakes and tasted like rubber. Snapshots of wives and sweethearts were examined with mutual appreciation; without benefit of interpreters, but using sign language, they talked far into the night and felt better men for it.

[3]

It would not be strictly true to say that Chinese forces, thinly disguised as 'volunteers', came into the Korean War without any warning: in early October 1950, prior to the crossing of the 38th Parallel by the United Nations Forces, Peking radio made a statement to the effect that the Chinese people 'would not stand idly by' while their North Korean neighbours were invaded.

The entry of sizeable Chinese units into Korea started on or about October 14th, and by the 24th three armies had crossed the Yalu. These were followed almost immediately by seven more armies, while a further three armies in Manchuria moved up to the border.

But the sinister build-up of Chinese manpower had originally started as far back as July, when it was estimated that there were 116,000 regular troops in Manchuria; by the end of August this figure had risen to a quarter of a million. September and October saw still more large-scale troop movements into Manchuria from central and southern China, so that by mid-November the strength of the Chinese Communist Forces in Manchuria and Korea was estimated at 850,000.

At least 500,000 of these were front-line troops: the Chinese soldier does not require the amenities which his Western counterpart regards as necessities, and almost every man is a fighting soldier; many of the non-combatant duties – particularly porterage – were performed by civilian labour, either locally enlisted or impressed. In such matters as rations, medical attention, communications, and the evacuation of casualties – duties which cause the administrative 'tail' of British and American divisions to be embarrassingly large – Chinese army methods were primitive in the extreme: they fought and marched on a bag of rice a day; wounds were patched up in a hit-or-miss manner or ignored altogether; for communications in the field they relied largely on bugles; only the higher ranking casualties were reported by name.

Soldiers of the Chinese Red Army were mostly illiterate peasants; physically enormously fit and hard and prepared to undergo any manner of discomfort without complaint. From earliest childhood they had been brought up to war. All their lives they had been beset by one burning problem: how to get enough to eat to keep themselves alive for one day longer. They had seen their wives, mothers, and sisters raped by marauding Japanese and Chinese Nationalist soldiers. They saw sudden, horrible, and obscene death every day of their lives.

A Communist Chinese life was cheap; there were no pensions, War Graves Commissions, convalescent camps, cinema shows, compassionate leave, concert parties, medical comforts, or toasts to fallen comrades. They went into the attack in hordes, and the amount of casualties was a matter of supreme indifference to the higher command. They continued to press home attacks even when their dead were piled high round the enemy's position.

This, then, was the new enemy . . .

[4]

Now it was a different war altogether, and it started disastrously. In the command posts of 8th Army and its subordinate organizations there were scenes of hectic activity and confusion as reports of fresh crippling reverses came in. The United States Army, almost overnight, had gone in a single short stride from verve and the fervour of victory to defeatism and near panic: once the succession of calamities got under way, there seemed to be no way of stopping them.

The South Koreans on the American right flank during the drive north from Pyongyang had run into stiff resistance: the resultant Chinese counter-attack stopped them in their tracks and finally crumpled and routed them. The 1st United States Cavalry Division, already earmarked for return to Japan and almost to a man disbelieving and sceptical about the rumoured Chinese intervention, were rushed to their assistance. On November 1st, in the area of the small town of Unsan, occurred the first big clash between Chinese and American troops: for the 1st Cavalry Division it was an unmitigated disaster.

The cavalrymen were encircled by hordes of Chinese – the word 'horde', loosely used in the early stages of the war, now had a very real meaning. The Chinese were aggressive and mobile; the Americans, with their usual embarrassing superfluity of vehicles were neither. In the small hours of November 2nd the evacuation of the 1st Cavalry began: vehicles, in many cases bumper to bumper, crowded for the last available inch on the ancient road which for years had only seen pedestrians and animals. From the mountains on either side of the road the Chinese fell upon the struggling convoy with savage and efficient fury.

The road was narrow, with a deep ditch on one side and a sheer drop on the other: vehicles went over the edge and into the ditch. There were stalled vehicles and burning vehicles blocking the road; unharmed vehicles – after the first three minutes of the action they represented a tiny majority – could not detour around the wrecks. And all the time the Chinese, who by this time had begun to laugh, cheer, and blow derisive bugles, poured automatic fire and hand grenades into the close-packed vehicles.

The 2nd United States Division fared even worse: they came down the twisting, turning road from Kunu-ri in the same tightly packed and defenceless formation – vehicles nose-to-tail, double-banked; now appearing on an open bend, now vanishing behind the dropping shoulder of a hill. And this convoy, too, took terrible punishment in that narrow enfiladed pass: machine-gun fire poured into the trucks; men fell out of the vehicles and were crushed into the frozen dust.

Everywhere the melancholy scene was the same, and for the hapless, road-bound Americans it became a dreadful war of ambuscade and stealth: the Chinese hid by day and attacked by night – fast-moving and lightly-equipped, they swooped on convoys under cover of darkness, struck sharply and murderously and then drew back to disperse in daylight.

American and Australian pilots performed prodigies of valour in their efforts to arrest this slaughter: they darted into narrow valleys to smash vital bridges and roared down almost to ground level, blasting the enemy out of earthworks and trenches. But air strikes were not enough; the terrain did not favour air power, and this the Chinese knew only too well.

It was retreat all along the United Nations front; if the northward advance had been precipitate, the November withdrawal was doubly so – this was the biggest and most humiliating 'bug-out' of the war to date.

After the triumphant advance to the Yalu, with the Chinese intervention there were ominous signs that the Army had become re-infected with the 'bug-out fever' of July. As one disillusioned GI, clearly not a career soldier, remarked: 'As soon as we see one little slant-eyed face, we shag-ass to hell out of it.'

But there was one American who never uttered the word retreat, and forbade its use in his division. Major-General Oliver Prince Smith was a marine to the bone, and in common with all United States Marines had a low opinion of the United States Army. The 5th Marine Division, he told a reporter whose probing had reached an infuriating level of persistence, were coming out as a Marine Division, bringing out their equipment, their wounded, and their dead. They were not retreating, withdrawing, 'bugging-out', or any other goddamn thing – they were merely fighting in another direction.

American ignorance of withdrawal tactics led General Walton Walker to have a discussion of frightening frankness with Brigadier Aubrey Coad. The arts of defence, withdrawals, and rearguards, Walker explained, were not taught at West Point or at the Infantry School at Fort Benning. Aubrey Coad, never a man to mince his words, said: 'When it came to withdrawal after the Chinese came in, the Americans had no idea what to do at all'; and he added wryly: 'but I soon caught on, and realized what his conversation was leading up to – 27 Brigade was to do rearguard.'

Coad's orders to Colonels Man, Neilson, and Fergusson were simple: 'You will go to Sunchon. If the road is clear, you will march there; if not, you will fight your way there.' The Argylls, Australians, and Middlesex eyed their allies with rough compassion, made a final brew of tea, shouldered their packs, swore horribly, and made ready to cover the retreat. They set off down the freezing road with the pipes of the Argylls playing *Cock o' the North* with shrill defiance.

But in spite of their outward air of philosophical resignation, the men of 27 Brigade were puzzled and angry: although they were retreating under orders, and appreciated the necessity for these orders, they did not know *why* they were retreating; they did not consider that the enemy had outwitted or worsted them at any stage.

The men of the Turkish Brigade were doing no retreating: like the Americans, they had not learned such a manoeuvre; unlike the Americans, they were not going to start learning it now.

The Turks could not retreat, even if they had wanted to: on their twenty-third day in Korea they had been hastily flung in to stem the Chinese onslaught at Tokchon; the South Koreans had crumbled, and the withdrawal of the 2nd United States Division could not, however far and charitably one stretched military phraseology, be described as 'orderly'. The Turks were encircled by Chinese of at least divisional strength – first they were subjected to shattering mortar fire, and then the bugles sounded and waves of Chinese infantrymen came at them from every side.

The Turks fought back with grim ardour: in a succession of bayonet charges – and the Chinese were nowhere near their match when it came to cold steel – some hacked their way through to the safety of the neighbouring hills; when their ammunition ran out, they continued to fight with bayonets, clubbed rifles, knives, boots, fists, and stones.

Conflicting reports came in regarding the fate of the Turkish Brigade; originally the 4,000 strong force was thought to have been completely wiped out, but casualties in killed and missing were eventually calculated at 1,000 as fresh parties of exhausted but indomitable Turks struggled back to the United Nations lines – they did not ask for a 'cushy' period in reserve; they asked for nothing save the chance to fight again.

One Turkish battalion refused to be rescued: the commander sent back word that they would fight their way out, and fight their way out they did.

The Turks fought with a ferocity which made them a legend in Korea; nothing, it seemed, had changed since Gallipoli and Mesopotamia: in one action they complained bitterly that an artillery barrage put in to soften up an enemy position was too heavy – when they reached the objective there were not enough live Chinese left to make it a respectable fight; in another, General Tashin Yazici told his men that the Chinese holding a ridge were the same ones that had cut them up so badly at Tokchon: the slope in question was about four hundred yards high, and had a gradient of about one in six. But the Turks swarmed up it and killed two hundred Chinese, mostly with the bayonet. On yet another occasion the Chinese ran before the final Turkish bayonet charge could be launched. 'It was very irritating,' said a Turkish officer diffidently to a newspaper correspondent; 'we had to put our bayonets back in their scabbards.'

Turkish casualties were almost all killed in action, missing, or seriously wounded; there were virtually no such categories as 'slightly wounded' or 'walking wounded'; slight wounds were usually ignored, and if a Turkish soldier could walk then he could fight. American and Indian medical orderlies reported that many wounded Turks refused aid, and several of them treated bullet wounds by dabbing mercurochrome over the point of entrance.

The durability of the Turks amazed everyone with whom they came into contact, and no one more so than Second-Lieutenant Stanley Fothergill of the Middlesex.

Fothergill's platoon were enduring all the misery of rearguard to a major retreat and its attendant discomforts: the temperature was 20 degrees below; they had not enjoyed a full night's sleep or a hot meal for a week; they were stuck out on a hill, seemingly a thousand miles from nowhere with, so far as Fothergill could ascertain from the scanty intelligence at his disposal, the entire Chinese Army hot on their heels.

Stanley Fothergill's orders followed the usual depressing pattern: 50 per cent stand-to all night; no smoking, no talking, no fires; the position will be held to the last man and the last round.

The Platoon Sergeant, Tom Heaffy, had tersely explained to the men that they were to shoot anything that moved and ask questions afterwards. 'The only thing that'll move out there,' he declared, 'will be a gook, or a pig, and if it's a pig for Christ's sake don't miss it': Heaffy spoke feelingly on the subject of pigs – only the night before Mr Fothergill had bagged a pig with his rifle: a pig of wonderful, juicy succulence on which the platoon had feasted with unashamed greed. Heaffy, who had only eaten a small tin of canned hash since, wanted a return to that golden age with plenty of crisp crackling.

Towards dawn voices were heard and the sound of shuffling feet. The bren gunner, a highly strung youth who also had a weakness for roast pork, fired a long burst. A prayerful howl of protest came from the darkness.

'What did he say?' asked the Number Two on the bren, passing up another magazine.

'Sounded like Burke,' said the bren-gunner, pulling back the cocking handle again; 'and he'd have to be one to wander about there on a night like this.'

But Stanley Fothergill forestalled a second burst of fire: he, too, had heard the cry – he knew no Chinese, but he had heard the voice. 'Come this way!' he shouted, and two shadowy figures warily approached the platoon position.

They were two privates of the Turkish Brigade, which had been all but decimated at Tokchon, and they were both in the uttermost extremes of exhaustion. One of them had an ugly bullet wound in the neck, which he had plugged with gauze and bandaged with a dirty rag. Simultaneously they both said: 'Turk.'

'What did I tell you?' said the bren gunner triumphantly. 'I knew they wasn't gooks.'

The two Turks were absorbed into the platoon, and by common consent were christened 'Abdul One' and 'Abdul Two'. 'Abdul One', running true to form, politely declined medical attention, but accepted an American cigarette with alacrity; 'Abdul Two', as if in apology for not being wounded, indicated by sign language that he was prepared to do sentry duty for indefinite periods.

As the weary slog southwards continued the next day, 'Abdul One' and 'Abdul Two' continued their attachment to Number Three Platoon: they were unquenchably cheerful and acknowledged no fatigue; in exchange for a cigarette, 'Abdul Two' volunteered to carry the bren gun for the next ten miles . . .

The men of the 16th Indian Field Ambulance did not take kindly to retreat, but theirs was the objection of the healer rather than of the fighting man. At Pyongyang 'bug-out fever' was at its height: the Americans had set fire to their vast supplies of petrol, clothing, and ammunition – on all sides flames crackled, and a choking pall of smoke obscured the scene of expensive wreckage; a veritable tax-payer's nightmare.

The Field Ambulance, recently arrived in Pyongyang, was caught up with peremptory orders to pull out. To Colonel Rangaraj, accustomed to a more frugal system of accounting, this course of action was unthinkable, and it was here that the extraordinary devotion and versatility of this Ambulance unit came into full play.

Standing beleaguered and apparently unwanted in a small siding was a dilapidated railway engine and a few trucks. It is unlikely that any medical unit of any army carries engine drivers on its strength, but 16th Indian Field Ambulance did; soon the men had formed a chain, and the precious stores were being loaded: a softly spoken Bengali, who had once been a fireman on the Deccan Mail (despairing of promotion to driver, he had enlisted in the army instead) looked the engine over with a professional eye, and reported to his Colonel that given fuel and considerable assistance from God it could be coaxed into motion. There was no coal, but wood in plenty. A chain of Indians passed up jerry-cans of water and the stores were loaded on to the trucks.

Then there was a violent hissing of steam, a triumphant shriek from the engine's whistle; the train with its precious cargo, manned by men dedicated to the task of alleviating human suffering, gathered speed and rolled through and out up the line, southward. American engineers, detailed to demolish the last bridge, stared incredulously and then cheered. It would be difficult to imagine finer initiative and devotion to duty than that lavished on the last train out of Pyongyang.

Now there was a double tide rolling south: the retreating army and the wailing, frozen refugees – long columns of ragged and starving scarecrows, surely the most pathetic and beastly aspect of war.

[5]

There was now a new British brigade in Korea, commanded by Brigadier Tom Brodie – a small, immaculate, restless man full of nervous energy: originally of the Cheshire Regiment, he had been a Chindit under Wingate and knew all about all the dirtier aspects of war. On arrival Brodie sniffed the air eagerly, like a restless retriever: 'Korea?' he exclaimed; 'what's wrong with Korea? I like it – reminds me of Scotland.'

NINE

Brigadier Aubrey Coad's 27 Brigade went from Hong Kong to Korea with virtually nothing except the 1st Argylls and the 1st Middlesex. Until the arrival of the New Zealand Artillery Regiment, they had no artillery support; they had no tanks or engineers; no medical unit, and only a bare minimum of atrophied vehicles. Not for nothing were they known as the 'Woolworth Brigade'.

At first sight Brigadier Tom Brodie's 29 Brigade seemed to have everything: three battalions of infantry and a cavalry regiment, all with illustrious titles and traditions to warm any military historian's heart – 1st Battalion Royal Northumberland Fusiliers (the 5th Fusiliers), 1st Battalion The Gloucestershire Regiment, 1st Battalion The Royal Ulster Rifles, and the 8th King's Royal Irish Hussars (this regiment had charged at Balaclava as part of the Light Brigade). 29 Brigade also had the 45th Field Regiment, Royal Artillery, the 170th Independent Mortar Battery, and the 55th Field Squadron, Royal Engineers, in addition to its own medical and vehicle repair units. For cleanliness within it had chaplains of every religious denomination; for outward purity it had a mobile bath unit and a mobile laundry unit. It had many highly qualified staff officers, numerous signallers, an ambitious-sounding organization known as an Aerial Photographic Interpretation Section, universally and ribaldly referred to as 'A-Piss', a detachment of the Royal Army Pay Corps, and its own military policemen.

But a second and longer look at 29 Brigade in England in the late summer of 1950 revealed that, despite its impressive strength on paper, it had hardly anything at all. For it lacked the one commodity that really counted in war – particularly in a war against the Chinese – and that was men.

The Brigade was also alarmingly scattered: Brigade Headquarters, 45th Field Regiment, the Gloucesters, and the Royal Ulster Rifles were in Colchester: the Gloucesters had recently returned from a sybaritic and sun-kissed tour of duty in Jamaica, and the Ulsters were just back from Germany. The 8th Hussars were in Tidworth, and the Royal Northumberland Fusiliers at the School of Infantry, Warminster, as demonstration battalion.

Every unit of 29 Brigade was, at best, at half strength: the infantry battalions were upwards of four hundred men short; the 8th Hussars could barely muster a full squadron; the gunners and sappers were in like case. Like the original 27 Brigade in Hong Kong (but by now Aubrey Coad's men were as battle-hardened as any formation of any nationality in Korea) almost every man of 29 Brigade, except the older officers, warrant officers, and more senior sergeants, would be going to war for the first time. Like the infantry battalions in Hong Kong, they were a mixture – in approximately equal proportions – of young regulars and National Servicemen.

In order to combat this depressing lack of manpower, the War Office had perforce to call out that variegated, griping, and peerless host – the Regular Army Reserve; in this respect Brodie's brigade came to resemble the 'Contemptible Little Army' of August 1914, which had been composed of approximately 60 per cent reservists and still saved Britain.

Of the Korea reservists, it has been written, not entirely truthfully, that they went to war without fuss or recrimination. Among the 'seven and five' men, yanked back to the Army in that summer of 1950, there was fuss and recrimination in plenty: as Regimental Sergeant-Major Jack Hobbs of the 1st Gloucesters neatly phrased it, 'they were ticking like alarm clocks'.

Not unnaturally, the married men with children – something like 75 per cent of them – 'ticked' the loudest: they were taking a steep financial dive from something like £10 to £15 a week (and in some cases considerably more) to £2 9s; their wives, who had gaily embarked on hire-purchase commitments for furniture, television sets, radiograms, and washing machines, abruptly found themselves in receipt of £4 10s per week. Meanwhile, there were mortgage payments and insurance premiums to be paid and children to be fed and clothed. It was scarcely surprising, therefore, that not every reservist recalled to the colours was burning with patriotism – in spite of their medal ribbons which spanned most of the universe, the majority of the reservists had never heard of Korea, nor did they want to.

They went back because, as they lamely explained to their dependants, they had signed on for 'seven and five' and had no choice in the matter: orders were orders, and it was back to the Army at the double.

The reservists flocked into Colchester, Bury St Edmunds (whither the 5th Fusiliers had moved from Warminster), and Tidworth, in audibly discontented droves, and their average age was 35; infantrymen for the Gloucesters, 5th Fusiliers, and Royal Ulster Rifles; tankmen for the 8th Hussars; gunners, sappers, and RASC drivers. There were Dunkirk veterans and Desert Rats; men who had stormed the Gothic Line and landed on D-Day; men who had dropped on to the Germans at Arnhem and into the sea at Sicily; there were men who had halted the Japanese thrust at Kohima and men who had been Chindits; men, captured at Dunkirk, who had spent five years behind barbed wire and others who had given three-and-a-half years of their lives to the building of the Burma–Siam railway.

Some had served with distinction and some without. Some had conduct sheets of pristine innocence, while others boasted documents running into many pages, covering insubordination, dirty kit and flesh, alcoholic and amatory indiscretions, and all possible variations of conduct to the prejudice of good order and military discipline. There were men who had been promoted and as speedily demoted; there were men with character 'exemplary' and 'excellent' on discharge, men with 'good' (a rating which smacks of mediocrity), and men with 'indifferent' and 'bad'. There were men who had volunteered for every sort of hazardous mission, and men who had never volunteered for anything except pay parade and weren't going to start now.

There were, of course, women involved. In this respect 29 Brigade differed widely from 27, for whereas 27 were predominantly bachelor, 29, in their reconstituted form, posed welfare problems the like of which the British Army has never encountered before or since.

There were shrill-voiced and belligerent women. Captain

Hugh Hamill,* then Adjutant of the 1st Royal Ulster Rifles, remembers a domestic incident which even today causes him to break out into a cold sweat.

A reservist's wife – a fierce young Liverpool-Irish matron – stormed into his office, having peremptorily shouldered her way past a regimental policeman, two orderly-room clerks, and – final indignity – the Orderly Room Quartermaster Sergeant. With her were two children, toddlers aged two and three, and these victimized mites were thrust on to Hamill's desk, which bore the clutter of papers which makes the life of an Adjutant – particularly an Adjutant caught up in the throes of mobilization – a penance of the most daunting description.

Pausing but briefly to draw breath, but not for long enough to risk interruption, the lady, having deposited the two unfortunate children, announced with strident authority: 'Right, mister, them two are yours from now on,' and flounced out of the office, leaving a spilled fire-bucket in her furious wake. The Adjutant was the object of wide-eyed and thumb-sucking scrutiny from the two infants, one of whom had overturned an ink pot and was now sitting in the ink; the other, not to be outdone, pelted him with paper clips.

There were, not surprisingly, a number of pregnant women. Colour-Sergeant Reg Panting of the 1st Gloucesters had been a father twice before, but he was not to see his new daughter until she was three years old. As the *Empire Halladale*, carrying the Royal Northumberland Fusiliers, approached Pusan, Fusiliers Sydney Shaw and Bill Chambers became the fathers of a son and a daughter respectively.

There were hopeful women, engaged to be married to men who, military service well behind them (or so they thought), were advancing gingerly towards the hazardous minefield of matrimony. Some, like 5th Fusilier Ted Fryett and Privates Ron Thomas and Don Chamberlain of the Gloucesters, rushed hastily to clergymen and demanded that the banns be read without further ado; less impulsive swains, like Fusilier Fred Pipkins, a Barnsley bricklayer who had fought from El Alamein to

* Now Lieutenant-Colonel Hugh Hamill, Commanding Officer of the same battalion.

the upper reaches of the Rhine, sought refuge in a more elastic courtship conducted at extreme range.

Many Royal Ulster Rifles reservists, private-enterprise enthusiasts, had started one-man businesses with their gratuities: Rifleman Tom Dorman, with a background of service spent as a transport driver during the war, had acquired a single car and converted it into a taxi – he confidently expected to own a fleet of six within a year, and probably would have. Rifleman Bill Graver fancied bigger vehicles, and in partnership with his brother-in-law had purchased a coach, which was much in demand for old-people's outings. Rifleman Arthur Tonks had opened a fish-and-chip shop which was, so he indignantly told an investigating officer, worth ten thousand pounds.

Rifleman 'Taffy' Williams, a Welsh miner, whose presence in the Royal Ulster Rifles had always been good for a laugh during the war years, had returned to the pit and found that his mining skill had become worth £18 per week: he had gone absent without leave three times during his not inconsiderable war service in North Africa, Italy, and Normandy, and on his recall was busily formulating plans for making it four. Rifleman Charles Fanning – veteran of Dunkirk, North Africa, Sicily, and Italy, married in June and recalled in August– had become a postman in Belfast; he had just finished a day's deliveries when he opened his own mail which consisted of one alarmingly significant telegram. Rifleman John McGuigan, wounded in Sicily and subsequently a fractious and uncooperative prisoner of the Germans for two years, had a row with the foreman one morning, received his call-up papers on the next, and flattened authority with a shrewd blow to the jaw on the same afternoon – he reported his arrival to Captain Hugh Hamill at Colchester with the encouraging announcement that he was thankful to be back in the Army ('a satisfied customer at last,' said Hugh Hamill, who had not yet fully recovered from the impact of the two children on his desk).

A startling note of originality was introduced by the appearance of Rifleman Red Martin who, with a beatific smile, informed Lieutenant-Colonel Robert Carson, Commanding Officer of the 1st Royal Ulster Rifles, that it was quite out of the question for him to fight against Communism in any form and produced a bewildering array of documents which proclaimed him to be a fully paid-up member of the Communist Party. Colonel Carson was visibly shaken, but gave no ground. 'You can always come along for the ride,' he told Martin reasonably, and an alarming roar from the RSM put paid to any further political discussion.

Rifleman Eric Pole, a police constable stationed in Farnham, Surrey, came off the early tour of duty ending at 2 PM and found his recall telegram waiting for him; for Pole it meant a swift change from one uniform to another. At almost exactly the same moment another uniformed reservist, bus conductor Ben Arless, severed connexions with the Belfast Bus Company.

Rifleman Patrick Clancy (occupation 'self-employed') very nearly got away with it – but not quite: on the eve of the troopship *Empire Pride*'s departure from Liverpool, he had confidently informed Second-Lieutenant Mervyn McCord that the ship would sail without him. On board ship McCord discovered to his dismay that this prophecy seemed to be coming true – all his platoon were present with the notable exception of Clancy.

But Clancy pushed his luck a little too far. On the quayside, just before the *Empire Pride* sailed, the all-seeing eye of Company Sergeant-Major McConville spotted him in the crowd. Soon two agile regimental policemen were trotting down the gangway, and closed in on Clancy; volubly protesting at what he considered to be a 'terrible crime against a neutral Irishman', he was taken aboard and became part of Liverpool-Irish history. An unwilling warrior indeed, but he and Red Martin were to fight well later . . .

The Army took the reservists back gleefully and fastened on them avidly: it medically examined them, took out bad teeth, cut their hair to spartan brevity, and encased them in hairy shirts, and battledress and heavy, nailed boots – items of clothing which they had long since consigned to dim memory. It taught them to drill again, and reminded them that officers must be saluted on all occasions, even when wearing civilian clothes; it sent them on route marches and inspected their feet afterwards; clothed them in vests and blue shorts and put them to rope climbing and press-ups; it brushed them up on the bren gun and bayonet; it sent them on rifle ranges and made them throw first dummy and then live grenades.

Officers of the 8th Hussars worked round the clock, categorizing drivers, gunners, and wireless operators. Old friendships were renewed, and in some cases old enmities. Men whom the reservists had remembered as privates had become platoon sergeants. Company Sergeant-Majors peered long and anxiously at faces which they piously hoped they had seen for the last time, and went into anxious huddles with company commanders:

Regimental Sergeant-Major Jack Hobbs, twenty-five years a Gloucester (five of them as a prisoner-of-war in Germany) emerged from his little cubby hole of an office adjoining the orderly room on a sunny August morning. From his cheese-cutter cap with its gleaming chin-strap to his glittering toecaps he was the epitome of military perfection. Hobbs, personally, was pleased enough about the pending move to Korea, but he was not given to analysing his feelings about any posting, whether it was to Tidworth or Timbuctoo.

A glance at the orderly-room clock told the RSM that it was twenty minutes to one, and by the same token time for a pint.

On his way across the square to the sergeants' and warrant officers' mess, he reprimanded two men for sloppy salutes and recommended an early visit to the barber's shop for two more.

He eyed the scene in the mess with satisfaction, and indeed it seemed traditional and everlasting: giving serious concentration to their beer were CSM John Morton of 'B' Company and CSM Mike Gallagher of 'A'; Colour-Sergeant Reg Panting and Peglar, the Provost Sergeant; 'Drummie' Buss, the Drum-Major and 'Knocker' Brisland, the medical sergeant – familiar faces, the faces of old friends, reliable faces. But in a corner could be heard a voice, and it sounded discontented. The RSM's eyes narrowed: the reservists were 'ticking', and Jack Hobbs sympathized with them (not that any of them would have thought so). But here was a *sergeant* 'ticking'; this would not do, not in the 1st Gloucesters; here was a shady affair.

The 'ticker' was a young sergeant, recently promoted, still more recently married. The gist of his complaint was that at long last he had a married quarter in a convivial home station, and that he had no wish to fight in a war which no one wanted any part of in a country which no one had ever heard of. It was unfortunate that he chose to deliver this diatribe as the RSM came into the mess.

'That's no way for a regular soldier to talk,' admonished Hobbs sternly.

'No, sir,' said the young sergeant; he couldn't really say anything else.

'Regular soldiers go where they're sent and don't moan about it, follow?' said Hobbs.

'Yessir.'

'Then don't let me hear any more bull like that again.' Hobb's stern features relaxed into a smile; he shouldn't really be reprimanding a fellow member of the mess at beer time, but a ticking sergeant . . . 'Give this horrible man a pint,' he said to the steward, an elderly man in a white coat who had finished with war many years ago, 'and put another one in there . . .'

In the main the men of 29 Brigade accepted the news with the same stoicism that their forefathers had displayed when ordered to the Crimea, the Sudan, South Africa, and Gallipoli. In any event, a good moan was a soldier's privilege – as long as it wasn't in the hearing of RSM Hobbs.

The 'ticking' of the reservists, like run-down alarm clocks, was gradually stopping: the men who had suffered extreme domestic hardship had gone home, having been investigated by special welfare tribunals set up to investigate each case; the remainder said 'lucky bastards', and grimly set about the distasteful business of becoming soldiers again.

There were something like two thousand of them in 29 Brigade, and almost as many women left behind who wept for them, pined for them, prayed for them, or were glad to see the back of them.

But once on the troopships the 'ticking' abruptly stopped and for good, for there was clearly no future in it: they were back in the Army again, and they had to make the best of it – if they didn't like it, then they would have to do the other thing.

In this manner the 29 Brigade, a mixture of green men and seasoned men, sailed for Korea.

TEN

WHEN the troopships carrying 29 Brigade sailed in the first week of October 1950, it seemed that the Korean War was virtually over: the North Korean armies had been shattered and broken and hurled back to the Yalu River; there might be a little mopping-up to be done, possibly some anti-guerilla work. But occupation-force duties seemed most likely, and Brigadier Tom Brodie was formulating plans for the acquisition of polo ponies from Manchuria. One thing was virtually certain: they would be home for Christmas.

The War Minister, Mr John Strachey, visited each troopship in turn: the Gloucesters in the *Empire Windrush*, the Royal Ulster Rifles in the *Empire Pride*, the Royal Northumberland Fusiliers in the *Empire Halladale*, the 8th Hussars in the *Empire Fowey*. He reminded the men that they were about to take part in a United Nations undertaking, something new in the annals of the British Army; their mission would be an important one for the future peace of the world. The reservists listened sombrely and suspiciously, and decided that perhaps this Korea caper wouldn't be so bad after all – at least they wouldn't get bombed or torpedoed on the way out.

There was to be no carefree loafing in the sun on this trip. There is something physically and mentally stultifying about a long sea voyage with nothing to do, and all commanding officers were determined that their men would arrive in Korea fit and ready for immediate battle if need be.

There was weapon training and physical training every day. And Saturday afternoons were given over to the traditional amusement of the British soldier at sea, 'housey-housey'.

The average British soldier was (and is) no expert on the ramifications of Communism, and the men of 29 Brigade were no exception, except for that staunch and fully-paid-up member of the Party, Rifleman Red Martin, who still regarded his presence back in the Army as a gigantic capitalist confidence trick.

Captain James Majury of the Royal Ulster Rifles found himself detailed to give a lecture on Communist Aims to Headquarter Company, and looked forward to the chore with the minimum of pleasure.

Looking at the serried rows of faces, all clearly prepared to be bored, Majury viewed the assignment with even less enthusiasm. He was, however, relieved that Rifleman Martin was not in the audience. For the riflemen it was just another lecture – preferable to weapon training or PT, but less entertaining than housey-housey or doing nothing.

Armed with the Army Bureau of Current Affairs booklet and a sheaf of notes, Majury embarked on his lecture: 'Communist parties are groups organized for political action designed to inaugurate a world-wide Communist social order by seizing and exploiting political power and economic resources in conformity with a pattern introduced in Russia in 1917 . . . strongly-held beliefs traced in origin to the doctrines of Karl Marx . . . the superior productivity of a state-planned economic system over private enterprise . . . Communists sought power through violent revolutions . . . In 1949 the Communists in China proclaimed the People's Republic of China . . . Mao Tse-tung . . .'

'WAKE UP AT THE BACK THERE!' roared Company Sergeant-Major Sam Henry, better known as Al Somers, the boxer; half a dozen nodding heads and a muted snore proved conclusively that the magic of ABCA was not doing its appointed work.

'Any questions so far?' inquired Majury brightly.

'Yes, sorr,' said Rifleman Callaghan. 'May we smoke please, sorr?'

Majury started in heavily on Moscow influence and Joseph Stalin, and then decided that he was getting nowhere: he would pack up the lecture, and throw the subject open to the house for argument and lively discussion. He was, however, forestalled by an earnest bespectacled lance-corporal who worked as a clerk in the orderly room. 'Sir,' said this young man, 'what do you think about dialectical materialism?'

James Majury shot a hunted look at CSM Henry. The Ser-

Argyll and Sutherland Highlanders in Korea returning from hill positions they had occupied, preparatory to advancing again

Gloucesters on patrol to locate the enemy

Stretcher bearers with Australia's 3rd Battalion taking casualty over frozen hills to jeep ambulance and casualty centre

'Somewhere about the 38th Parallel' the 3rd Battalion The Royal Australian Regiment prepare to attack Sardine Hill

The 3rd Battalion The Royal Australian Regiment resting, having chased the Communists out of this vantage point

Lt-Col James Carne, VC, DSO

Private William Speakman, VC

Lt Philip Curtis, VC, DCLI

Major Kenny Muir, VC

The Hook, behind the line in Korea, saw more enemy activity than any other Commonwealth front line position

Hill 235 where Lt-Col Carne (Gloucesters) ordered his bugler blow
the long reveille defying attackers in epic last stand

Vigilant Australian medium-machine-gun teams scanning hills for Communist activity

Battle of the Hook. *Left to right:* Privates Bill Morrison, Jim Walker, Jim Flynn, Tom Stanley and Piper D. Ramsey

Drum-Major P. Buss (Imjin Battle commander) in the second batch of prisoners of war leaving for the UK on *Empire Orwell*

geant-Major, getting the gist of the thing at once, barked out a single crisp order: 'Outside for weapon training!'

Afterwards he said to Majury: 'Diabolical materialism! What the hell next . . .?'

Rumours were sweeping the *Empire Halladale*, the *Empire Windrush*, the *Empire Fowey*, and the *Empire Pride* as the mountainous coastline of Korea came into sight, partly hidden by driving squalls of rain.

It was becoming clear that the whole character of the Korean War was changing and that a completely new phase of operations was beginning. In the officers' lounge of each ship a war map was maintained, and anxious-faced battalion-intelligence officers were busy with chinagraph pencils. New symbols appeared and a succession of ominous arrows representing attacking forces and retreating forces; the attacking forces were in massive preponderance, and the arrows were red.

It soon dawned on everyone from Brigadier Brodie to the youngest soldier of the Brigade – 18-year-old Private Albert Hawkins, who had successfully persuaded Captain Anthony Farrar-Hockley, Adjutant of the 1st Gloucesters, that he was 19 and had therefore laid himself open to charges of perjury and false enlistment* – that there would be considerable more to Korea than occupation duties, victory parades, and mopping-up. For this was going to be a war, and it would be every bit as vicious, bloody, and sordidly uncomfortable as anything that the years between 1939 and 1945 had offered.

[2]

Evidence of Chinese intervention in Korea was not immediately apparent: indeed, the men of 29 Brigade began to wonder if they would ever see a Chinese soldier. Invincible hordes of them were supposed to be sweeping down from the Yalu, but the only hordes on view were American, and they were clearly declining to put their invincibility to the test.

* The Adjutant would probably have sympathized. In 1939, Brigadier Anthony Farrar-Hockley, DSO, MBE, MC (to give him his present-day title), enlisted under age himself.

After the by now conventional greeting at Pusan, the Gloucesters, Royal Northumberland Fusiliers, and Royal Ulster Rifles embarked on a train journey of surpassing tedium and misery to Suwon, where the Brigade were to form up. Food was prepared on the platform or beside the rails when the train stopped for long enough and the staple diet for the journey consisted of the inevitable bully beef, dehydrated potatoes, and Army biscuit, which was good stuff if you had strong teeth. The young soldiers gazed with awed fascination at the panorama of ricefields and gaunt hills, while the reservists, with a more philistine regard for scenery, opined that Korea, in common with many other places they had visited during the war years, was not worth the trip.

They saw nothing of war, but plenty of its beastly and criminally wasteful aftermath: overturned trucks, burnt-out tanks, trains twisted and gutted by fire, bomb craters, and the dismal wreckage of masonry and rubble where buildings had once stood.

And ever and anon they saw the refugees, shuffling along in panic-stricken, frostbitten, wailing thousands.

At each station on this jolting and bone-shaking journey these wretched creatures, all but dehumanized by poverty and cold, crowded round the train and their prayerful wailing for food sounded like the howling of mad dogs in a pit; they fought one another for the scraps of biscuit and chocolate which the soldiers threw from the carriage windows.

Then the men of 29 Brigade were on the road in trucks driven by Negro drivers, speeding northward to a battle which was never fought. They saw no enemy and there were only lurid tales, gleaned from Americans who paused long enough in their retreat to talk to them, to substantiate his existence.

The absence of any enemy and the demoralizing atmosphere of headlong retreat were particularly galling to the 8th Hussars. The Hussars, with Cromwell and Centurion tanks instead of horses, but full of the heady spirit of the cavalry, searched hopefully for good tank country. There was, alas, little enough of it to be seen.

There was, in fact, very little of anything on view save the undignified spectacle of the American Army in precipitate wheelborne retreat – a dark, grey-green stream of military traffic:

trucks, tanks, jeeps, troop-carriers, guns, and half-track vehicles, seemingly inextricably jumbled together.

There seemed to be no semblance of traffic control: white-helmeted military police in jeeps screamed up and down the column, trying to restore order but frequently merely adding to the chaos; vehicles cast tracks and had to be abandoned; slow trucks broke down and came to a grinding and obstructive halt, while faster trucks tried to 'jump the column'; there were Negro soldiers with eyes rolling wildly, who made no secret of the fact that the retreat couldn't go too fast for them; there were tough-looking bearded soldiers, staring grimly in front of them and clutching their rifles, as if the Chinese army was just around the corner in front of them instead of a hundred miles to their rear; there were sleeping soldiers who had been in hard action, and waking soldiers who had been in no action at all and fervently hoped that they never would be.

At a roadside halt, BBC correspondent René Cutforth shared a bottle of Korean whisky, the type of liquor which removes French polish if spilled on a table, with a Texan lieutenant: the Texan needed a shot of liquor badly, and didn't care who knew it.

'Jesus!' said the lieutenant, after the first swallow, 'does this fire water kill?'

'I'm still alive,' Cutforth reassured him. 'Tell me, how far behind are the Chinese?'

'That's something nobody ever seems to know,' said the lieutenant. 'They just turn up in half-millions, any place, any time. All we do is haul-ass down this road.' He took another comprehensive swig from the bottle. 'Tell you one thing,' he went on, with a candour which was scarcely reassuring, 'this is a new kind of war and we don't know what in hell to do about it.'

This was all too depressingly apparent. The very smell of panic and defeatism was in the air. The 8th Hussars remember that retreat, but they do not eagerly summon back the memories. The 8th Hussars had taken part in other retreats – in fighting retreats, from which victory was eventually secured. But from this retreat it seemed that nothing could be salvaged but shame and disaster.

'The conduct of this campaign to date,' said Captain the Honourable Rowland Winn with austere euphemism, 'had been uninspiring.'

But amid the misery and the cold and a disillusion that was harsher than cold, there was a glimmer of sanity in a world which seemed to have gone mad: in Pyongyang, the men of 29 Brigade met those of 27 for the first time since their arrival in Korea.

Most of the Argylls and Middlesex had gone to Korea with the faces of eager boys, but they were now unquestionably men: their faces were no longer boyish, but deeply etched with lines – the faces of men who had experienced the uttermost extremes of exhaustion and endeavour.

The men of 29, the Argylls and Middlesex noticed, were like the 3rd Royal Australian Regiment – considerably older than they were.

There was trouble in plenty in Pyongyang. The Americans were pulling out with their accustomed *élan*, and were making sure that the advancing Chinese found no comforts in the city: clouds of smoke obscured the wreckage of burning dumps of petrol, clothing, and ammunition; the air was thick with flying sparks and cinders, and from time to time a shattering explosion announced the destruction of yet another ammunition dump. Unattended vehicles were everywhere: Second-Lieutenant Ian Rutherford of the Middlesex was dispatched into the city by his company commander with the terse instruction to come back with a jeep or not at all; Rutherford's only regret was that he could only drive one at a time.

In a dozen American stores depots the British soldiers got the same answer from harassed officers: 'Help yourself, mac – we're blowing the whole goddam place to hell in a half hour so nobody ain't gonna miss it.'

There were bales of warm winter clothing – brand-new-pile jackets, fur caps, socks, boots to set any quartermaster dreaming; there were rations by the ton, fruit juice by the gallon, and cigarettes by the million. There was scarcely a vehicle of 27 or 29 Brigade which was not laden to the roof.

In this manner, without sighting a single Chinese soldier, in the wake of a fleeing civilian populace and army, 27 and 29 Brigades went back – back to the inhospitable outskirts of Seoul.

[3]

The speed of the American retreat had completely outdistanced the Chinese advance, and gave the United Nations ample time to form a new defensive line – it ran from the coast near Pyongtaek, on the 37th Parallel, to near Wonju in central Korea, and thence to the east coast at Kangnung, some twenty miles south of the 38th Parallel.

On the extreme left of the line near Pyongtaek was 29 British Brigade, with the 3rd United States Division on its right; 27 was at Changhowon-Ni.

Suddenly it was Christmas Day, 1950.

By courtesy of the slowness of the Chinese advance (or the speed of the American retreat) and as a result of near miracles of improvisation, the men of 27 and 29 Brigades celebrated Christmas in near-traditional style in blinding snowstorms.

For RSM Hobbs of the Gloucesters Christmas was another military day. Like all RSMs, he spent many of his waking hours looking for men who were doing nothing without good and sufficient reason: on the afternoon of 24th December, 1950 he selected his victims with unerring accuracy, and took swift and characteristic action. Pointing an accurate finger at a hill, not more than half a mile of walking distance away, he bawled at three soldiers: 'Christmas tree, you three! Not tomorrow, but today! MOVE YOURSELVES!' The Gloucesters had their Christmas tree.

They also had a cocktail of startling potency. The sergeants, with memories of enterprising wassailing in Jamaica to guide them, produced a punch compounded of gin, whisky, rum, orange juice, American fruit cocktail, local hooch, and evaporated milk. Cook Corporal Taffy Watkins, who had cooked Christmas dinners on the North-West Frontier of India, in the desert in North Africa, and in Italy, displayed a staunch determination to be undeterred by the elements: turkeys, which were frozen solid on arrival, were thawed out and then roasted in ovens improvised out of petrol drums.

The 8th Hussars – the cavalry always strove to be a jump ahead of the 'feet' in matters of good living – prepared for Christmas with enormous ingenuity: the sumptuous table set up by Major Pat de Clermont and 'A' Squadron mess was a source of wonder to the luckier Americans who found themselves in the vicinity.

The Argylls were dubious about Christmas: it is not customary to celebrate this peculiarly Christian feast in a Presbyterian community. But Colonel Leslie Neilson had no idea where the battalion might be on New Year's Day, so he decided to range himself alongside numerical Sassenach superiority.

And so leaving the minimum number of sentries to watch for possible Chinese interruption, British soldiers trooped down from the icy mountains and filled their mess-tins with turkey, carved by the officers.

The menu had a distinctly American flavour about it, but there were no complaints: it consisted of tomato juice cocktail (contemptuously waved aside by the sergeants of the Gloucesters who had their punch) and relish tray. There was roast turkey in portions sufficiently gargantuan to satisfy any Briton – only the trimmings had a foreign appearance: celery dressing, giblet gravy, snowflake, and candied sweet potatoes, cranberry sauce, and whole kernel corn.

The Argylls and Royal Ulster Rifles had their bagpipes; the Gloucesters and Royal Northumberland Fusiliers explored the ruins of Seoul and came up with a piano apiece: battered and mildewed instruments with many notes missing and strings as gaunt as the frayed wishbones on a chicken. Carols were sung; unit comedians did impersonations of sergeant-majors; there were soldiers who sang, and soldiers who cried, and soldiers, overcome by crude alcohol, over-eating, cold, and acute homesickness, who vomited defiantly into the snow and came back for more. Climatically, it was as rough a Christmas as the British Army in the field has ever experienced, but they did their best with it and emerged the undoubted victors over the gross elements.

Then all too soon it was 1951, and Leslie Neilson's forebodings about New Year's Eve were amply confirmed . . .

The advance of 'the bottomless well of Chinese manpower' was inexorably catching up – not in trucks and tanks, but on foot, ox, cart, packhorse, sledge, and two-humped Bactrian camel; while it seemed that it was timed to give the Sassenachs a comparatively peaceful Christmas, it was also clearly designed to disrupt Hogmanay for the Argyll and Sutherland Highlanders. Nevertheless, the Argylls managed to achieve what they cautiously termed a 'wee bit party' with the Middlesex and 3rd Royal Australian Regiment, although the distant sound of gunfire joined in the sounds of revelry . . .

[4]

The year 1951 started badly.

In the United States Army, morale was at its lowest ebb: the retreat, following MacArthur's 'home for Christmas' forecast, the gruesome winter, and the newly created myth of the invincibility of the Chinese combined to produce a disastrous mental attitude. On all sides there was talk of retreating still farther and even of evacuation. 'Bug-out fever' still had the Americans firmly in its grip. Prognostications of disaster had reached such a pitch that the men of 29 Brigade were beginning to ask one another in all seriousness: 'What will happen to us when the Americans leave the country?'

But it would not be true to say that there was any atmosphere of defeatism in 29 Brigade; with Brigadier Tom Brodie in command it would not have survived for long in any event. Rather were they puzzled and ashamed of being involved in such a business: there were many Dunkirk veterans among them, as well as men who had been at Tobruk and Crete, Singapore and in the 1942 retreat out of Burma. But no one had ever experienced an operation like the recent withdrawal, which bore no possible resemblance to an operation of war. The reservists had fought and beaten Germans and Japanese, and were unconvinced that the Chinese were all that good. They wanted to go home, but throughout the Brigade there was a general desire to get down to business and discover the facts for themselves.

Perhaps this attitude of mind was best expressed by Lieutenant-Colonel Kingsley Foster, Commanding Officer of the 1st Northumberland Fusiliers who wrote to the regimental journal, the *St George's Gazette* as follows: 'We do not know when we shall be ordered home, and it would be hypocrisy to say that we do not wish that the day comes soon. But we are the King's men, and if his business is not satisfactorily accomplished, then there can be little joy in homecoming.'

This attitude was heartily approved and shared by a new United States general who had arrived in Korea in succession to General Walton Walker, tragically killed in a jeep accident on his way to present Presidential Citations to the 1st Middlesex and the 1st Argylls.

The new broom was Lieutenant-General Matthew Bunker Ridgway, and he promptly set about sweeping clean in every possible direction.

Montgomery had his richly assorted headgear, Patton his two pearl-handled revolvers, Wingate his beard and battered pith helmet. Ridgway was never seen anywhere without two primed hand grenades precariously attached to his chest, and he rode in a jeep fitted with an ear-splitting siren. As a result of his lethal breastplate, he was given the intriguing nickname 'Iron Tits'.

As commanding general of the 82nd United States airborne division, Ridgway parachuted with his soldiers into Normandy in 1944, and subsequently led the 18th airborne corps in action in Holland, Belgium, and Germany. Now he was in Korea, and there was not a soldier of any nationality or rank who was not made immediately aware of the fact.

There would, declared Ridgway forcefully, be no more 'bugging-out' (Walton Walker said the same thing many times, but not apparently with the requisite degree of force).

Matthew Ridgway was a tonic and a much-needed shot in the arm, and the effect of his doctrine on the Army was astringent and immediate; in a very short space of time he transformed what René Cutforth described as 'a mob of dispirited boobs, still thinking in terms of a press-button war' into a tough and aggressively minded force.

One, two, and even three-star generals, who had done no walking to date and did not intend to if they could help it, were instructed in no very ambiguous terms to get off their fat asses and get climbing hills; Ridgway joked with the soldiers, cajoled them and derided the faint-hearted; he established himself firmly in the affections of British, Canadian, and Australian soldiers. The misery of the retreat and the cold had taken the heart out of many American soldiers, but Matthew Ridgway put it back.

Ridgway was not interested in 'real estate'; he was unconcerned about ground lost and won. He thought simply in terms of inflicting as many casualties upon the Chinese as possible, and he impressed upon everyone that the enemy's only superiority was in numbers – 'all we got to do is kill more of their guys than they do of ours, it's as simple as that. And we're going to win this war – that's for sure . . .'

[5]

Brigadier Tom Brodie made frequent, and often unheralded, visits to his units – one of his more spectacular efforts was to land by helicopter on the Royal Ulster Rifles, missing by the narrowest possible margin the officers' latrine – and discovered that the reservists were settling in well. This did not greatly surprise him, but he had heard some of the early 'ticking' – some of it at embarrassingly close range – and was anxious to see for himself.

Brodie need not have worried, for the 'ticking' had stopped once and for all: admittedly, there were men in 'The Old Man's Brigade', as 29 Brigade was affectionately known, who were still more concerned with varicose veins, incipient arthritis, financial and family problems than with the prosecution of the Korean War, but they now kept it to themselves; they were concentrating, in fact, with the twin compulsions of army discipline and self-preservation to guide them, on staying alive. They knew all about war and its attendant discomforts, and weren't going to waste their energies in any unnecessary activity or noise.

From their experiences in North Africa, Italy, Burma, and France they knew the luxury of curling up in blankets on a groundsheet; they knew all about making rations last and digging shelters with the minimum of effort; they had learned how to live in vile climatic conditions and how to die without causing anyone else too much trouble.

It was, if anything, harder on the young regulars and the few National Service volunteers, for they were doing it for the first time. And to them the old hands proved fine mentors.

At first, back in Colchester, Corporal Holliday of the Ulsters, nicknamed 'Doc' after the famed alcoholic medical man of the Old West, eyed the additions to his section in the machine-gun platoon with awe and apprehension. Holliday had recently celebrated his twenty-first birthday; a keen and ambitious soldier, he had won his stripes early in his service and was already in line for accelerated promotion to sergeant.

But the three additions to his section – Rifleman 'Yammer' Youds (so called because he seldom stopped talking for long), Billy Reeves, and John Muir were something outside Holliday's ken: they were all aged about 35, all wore five medal ribbons or more; at first sight they were surly, belligerent, and relentlessly insubordinate.

Youds told his corporal that he had killed his first German when Holliday was still in short trousers (which was true); Reeves disobeyed two lawful commands on the second day, and Muir told him to get f— on the third.

On the troopships, and later in Korea, they all three abruptly ceased to be grousing civilians in uniform and became soldiers again; all three accorded Corporal Holliday a loyalty the like of which he has never experienced since. For once in the snowy wastes around Seoul they gave him the benefit of their variegated battle experience: they advised him on angles of sight, alternative positions, ammunition supply; and never for one moment did they imply that they should have been commanding the section.

[6]

The Chinese attacked all along the line in the small hours of January 3rd, and the main weight of 'hordes' – this word had, reasonably enough, come into usage again – fell on the 1st Royal Ulster Rifles and the 1st Royal Northumberland Fusiliers.

The action started badly for the Ulsters, and 'B' Company, commanded by Major John Mulligan, were enveloped by a carefully planned and skilfully executed attack: the Chinese advanced under cover of darkness close to 'B' Company's position, and at about 4 AM closed in on the Ulstermen, led by a man carrying a white flag and shouting in unison: 'South Koreans, we surrender!' to the accompaniment of a shrill chorus of bugles and whistles. Momentarily confused by such tactics, 'B' Company allowed them to come on, but it was not until the Chinese had effected a deep penetration of the company position that they realized their mistake.

The Chinese, elated by their initial success – a bugler, in quilted winter uniform and dog-skin cap, was with rare effrontery blowing a discordant success signal, and continued to do so until scared away by an irate machine gun – were not allowed to enjoy their superiority for long: Mulligan's men closed with the Chinese and for several confused minutes there was a desperate hand-to-hand skirmish; the Ulster Rifles do not like having such tricks played upon them and reacted with swift violence.

'Saucy lot of sods,' commented Major Hugh Gaffkin, when he heard of 'B' Company's misfortune, and then with Lieutenants Robert Benson and Colin MacNichol in the van, he took 'D' Company into a counter-attack with verve and fury which even the 3rd Royal Australian Regiment might have envied; in a swift mêlée of grenades, automatic fire at close range, and gory work with the bayonet the lost ground was speedily retaken.

As the murky and freezing dawn of January 3rd broke, the attackers were seen in precipitate retreat, but the Chinese were not giving in easily: in the valley ahead of the Ulsters' position reinforcements, estimated at something like one thousand strong, were massing for yet another attack.

Then ensued, in the words of Captain James Majury, the battalion mortar officer, 'a scene of splendid destruction': the closely packed Chinese, either displaying amazing courage or extraordinary stupidity, were mercilessly exposed to the full fire-power of the Royal Ulster Rifles: in addition to the mortars – Lieutenant Alan Hill's detachment were pumping bombs into them over open sights – there were the machine guns, and Rifleman 'Yammer' Youds kept his thumbs pressed exultantly to the firing lever to the accompaniment of a lurid running commentary.

A roving detachment of 8th Hussars' tanks, commanded by Captain Donald Astley-Cooper, not to be outdone, joined in with machine-gun fire. But it was left to the United States Air Force to administer the *coup de grâce*: with provocative ostentation, four Shooting Star jet fighters circled once low overhead and then screamed into the valley, hurling rockets, cannon shells, and napalm into the Chinese.

At midday Major Charles Blake, in temporary command in the absence of Colonel Carson who had been evacuated sick, was able to report to Brigadier Brodie that the Royal Ulster Rifles were back in their original positions.

'Good boys,' said Tom Brodie briskly to Ken Trevor, his Brigade Major; 'now, what about the Fusiliers?'

The 5th Fusiliers, in the area of Chunghung Dong, were doing fine. Three high hill-features were held by 'X', 'Y', and 'Z' Companies (the 5th Fusiliers, for some reason which is hidden in the mists of time, call their four rifle companies 'W', 'X', 'Y', and 'Z' instead of the usual 'A', 'B', 'C', and 'D'). Colonel Kingsley Foster gave this order to the company commanders, Majors Reggie Pratt, Robin Leith-MacGregor, and John Winn: 'Stand your ground, wait for 'em and hit 'em with everything you've got.' The Fusiliers did just that.

The Chinese swarmed up the almost sheer hills in screaming waves, and ran into the controlled and withering fire for which the 5th Fusiliers are rightly famous. 'X', 'Y', and 'Z' Companies stood firm, and the Chinese only succeeded in breaching the line at one point west of 'Y' Company; this infiltration was speedily dealt with by the reserve company, Major Charles Mitchell's

'W', and by 5 PM Kingsley Foster was able to report to Brigadier Brodie that all positions had been held.

The Royal Ulster Rifles were firm in their positions, and so were the Northumberland Fusiliers: the morale in both battalions was sky-high, because the Chinese, the enemy whose 'hordes' had become almost legendary in Korea, had been met,

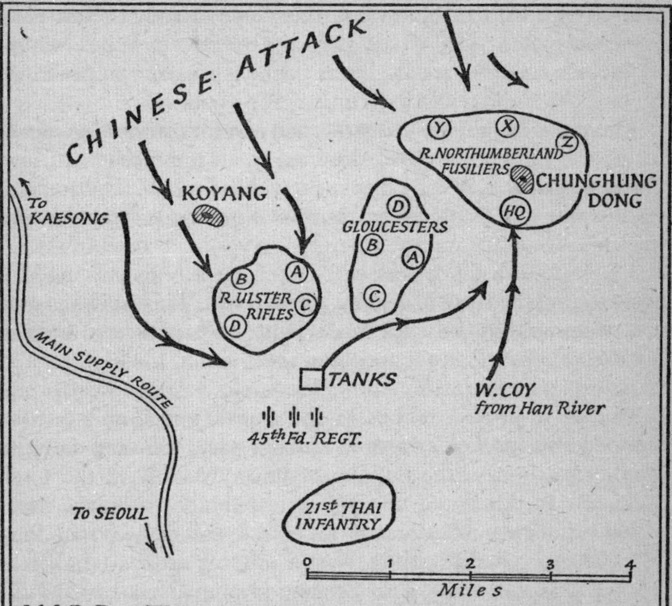

MAP D THE ACTION AT CHUNGHUNG DONG

© CASSELL & CO LTD 1967

fought and sharply and decisively defeated. Actions speak very much louder than words, and both the Ulsters and the 5th Fusiliers had something to crow about: a physical count inside the Rifles' position showed thirty enemy dead, and those killed by the mortars, machine gunners, and artillery must have amounted to at least one hundred more, at a cost of four killed and sixteen wounded; the 5th Fusiliers had sixteen killed and forty-five wounded, but jubilant statisticians in the battalion reckoned the Chinese dead on the slopes around Chunghung Dong at three hundred or more. And they had only just started . . .

The Royal Ulster Rifles were not best pleased to learn that they were to withdraw from the positions which they had held so tenaciously. This, however, was to be no 'bug-out', but a carefully planned withdrawal of the whole brigade to new positions south of the Han River, although to Captain James Majury the laconic announcement, 'We're gonna pull', from an American captain had a depressingly familiar ring about it.

The night of January 3rd was a bad one for clandestine movement: the temperature had dropped to an unprecedented low, and it was a night of inky blackness with no moon. The route of withdrawal was south by a pass over a low range of hills. The track was icebound and treacherous – a veritable sheet of glass – and for some distance was overlooked by a high curving cliff overhanging the river bed along which it ran. This pass was only just negotiable by vehicles in daylight; to experienced eyes in the Royal Ulster Rifles it appeared that, should the high ground be occupied by the enemy, the valley would become a death-trap.

And so it proved to be: as the trucks, tanks, and carriers bumped and lurched at a slow walking pace, grinding along in bottom gear and slithering almost uncontrollably in the pitch darkness, an American aircraft unaccountably dropped flares which brilliantly illuminated the black column of marching riflemen and vehicles against the surrounding snow. Major Blake sent a blistering request to Brigade Headquarters to have the illuminations stopped, but the damage was done: appallingly exposed, the column became a sitting target for a hail of mortar and machine-gun fire from the surrounding heights.

'B' and 'D' Companies escaped this ambush, but the remainder of the battalion and Captain Donald Astley-Cooper's roving troop of 8th Hussars were caught in deadly enfilade.

It was a scene of fearful slaughter: men crouching in the scant cover afforded by a shallow ditch at the side of the road were raked with machine-gun fire; men were shot off tanks and carriers; vehicles slid off the track in the confusion while other vehicles stalled or skidded to a halt.

Then the Chinese rushed down the hills, swarming down the slopes like a football crowd. They came from both sides of the track in a yelling horde, to the eerie accompaniment of whistles and bugles, and the trapped Ulsters fought back with grim ardour: their machine guns tore great gaps in the Chinese, but still they came on: they swarmed round the tanks and dragged men off; Corporal Sam McGrath machine gunned a group of Chinese at a range of about three feet and then, his ammunition exhausted, brained three more with the butt.

There were many acts of gallantry on that terrible night. The Chinese set up a machine-gun post from which they could spray the Ulstermen more easily, but Second-Lieutenant Mervyn McCord, with all the exuberance of his twenty years, collected a force of half a dozen men, and with Sergeant Harry Campbell and Corporal Bill Blackstock in the van grenaded it out of existence.

Some managed to break out: Major John Shaw led about fifty men in a bayonet charge and subsequently to safety; others managed to dodge the Chinese in the darkness and escaped; Corporal John Phillips crawled from under a pile of Chinese dead and marched alone for the rest of the night, eventually rejoining the battalion three days later – he had been a prisoner-of-war once, and once was enough for him.

For Captain James Majury and a party of some thirty men of Support Company there was clearly little chance of escape: the road in front of them was blocked by a broken-down carrier; the road behind them was swarming with Chinese, as were the hills on both sides. With grim despair Majury made a quick appreciation of the situation: if he shouted 'Every man for himself – run for it!' some of his men might have got clear, but this was not Majury's way, or indeed the way of any Royal Ulster rifleman. 'What'll we do, sorr?' inquired Rifleman Jack Orr, Majury's driver.

'Stay here and fight it out,' said Majury; he only gave one more order, and that was approximately thirty seconds later: 'LET 'EM HAVE IT!'

The Chinese came at them from all sides into a hail of bullets from the defiant riflemen; Majury himself emptied two complete sten-gun magazines into the advancing press of bodies; Rifleman Callaghan, in a fine frenzy of rage, dispatched four Chinese before falling to a bayonet thrust. When Majury's men were overwhelmed in the final rush, there were something like forty dead Chinese all round them.

Now they were prisoners, and to James Majury it seemed a poor end to eleven years of service. He had not been closely involved in the Second World War – the permutations of military postings had sent him to the North-West Frontier of India, and he had been pleased enough to go to Korea, even though he was just getting used to the idea of having a son, now a vociferous three months old; Majury had gone to the war in Korea with the simple desire to prosper in his military career.*

Later in a prisoner's compound, whither they had come after a dreadful forced march to the north, Majury was tortured by self-doubt: perhaps, after all, he should have ordered his men to run for it; perhaps the higher command would have thought that he had lost forty good effectives in pyrotechnics indulged in for personal ends.

He need not have worried. At the first Communist indoctrination lecture a stout Chinese, attended by a smirking interpreter with an alarming squint, told them about the 'lenient policy' of the Chinese People's Army towards prisoners: the British soldiers were 'dupes', 'tools', and 'enemies of the people'; they had been duped by the American imperialists and reactionary warmongers into fighting against the righteous cause of the Korean people, supported by their brothers the Chinese people. They were lackeys and hirelings of the barbarous Rhee puppet-government. The high-pitched voice went on, while Majury wryly thought that when it came to giving lectures on Communism this character could teach him a thing or two.

Now the Chinese officer was on the lenient policy again: 'Obey our rules and regulations and you will not be shot. You will be

* He has done so: he is now a Brigadier.

given the chance to study and correct your mistakes. You will—'

'Hey, mister!' shouted a rich Belfast voice.

The lecturer paused in mid-sentence. 'Yes, comrade?'

'Why don't yez go and f— yerself!' On hearing this piece of advice, all James Majury's doubts and recriminations abruptly vanished; this explicit rejection of Communism in Korea – it was one of the first of many uttered by captured British prisoners – proved once and for all to James Majury that he could stop indulging in self-reproach.

It was, indeed, a phrase repeatedly used by British Commonwealth soldiers to their captors – the Australians made even more original suggestions; less frequently by Americans, who have never shone over-brightly as prisoners-of-war. The many Britons who had the misfortune to be captured by the Chinese in Korea were ever present thorns in the sides of their captors, whom they reduced to exasperation and, eventually, unwilling admiration: the British prisoners flagrantly sabotaged Communist indoctrination with explicit trade-union efficiency, and greeted beatings and humiliation with the same impossibly ribald instruction which had come so spontaneously from an unnamed Ulster rifleman from Belfast.

One British prisoner-of-war, Fusilier Derek Kinne, did so with such good effect that he was awarded the George Cross. In the words of the official citation: 'Every method both physical and mental was employed by his captors to break his spirit, a task which proved utterly beyond their powers. Latterly he must have been fully aware that every time he flaunted his captors and showed openly his detestation of themselves and their methods he was risking his life. Nevertheless he was always determined to show that he was prepared neither to be intimidated nor cowed by brutal treatment at the hands of a barbarous enemy.'

[7]

Theorists on the vexed subject of Chinese invincibility could count on little support from the soldiers of the Commonwealth or other countries who had come to aid the United States. And in American units, too, since the arrival of General Ridgway, the phrase was losing much of its potency.

They all realized, of course, that fighting the Chinese was no pulp-magazine adventure, but a stark and deadly business: they were an enemy to be treated with respect in mobile warfare, particularly at night; they were tough and cunning; they were able to march enormous distances without any loss of efficiency, carrying enough rice, oatmeal, and dried peas or beans for ten days in cloth bandoliers slung across their bodies.

But the Chinese soldiers were not well armed: in a section of ten men, probably four or five carried nothing except grenades; their orders were to acquire further weapons from the dead. As an individual the Chinese soldier was far from intelligent – Ulster riflemen and 8th Hussars who survived the night of January 3rd–4th reported that many of them appeared drugged or drunk; certainly their reactions were slow, and they rarely got in the first shot at close quarters.

Man for man, the British soldier was his superior and in a straight fight they felt they could always win. And that went for Australians, Canadians, Turks, Greeks, Belgians, Puerto Ricans, New Zealanders, and all the strange miscellany of nationalities which made up the United Nations Army.

The Chinese invincible? Ask Gunner Bob Sprague, a slow-moving, slow-speaking but not slow-thinking farmworker from the Isle of Wight, who dispatched ten of them at close range with a bren gun.

Ask the Greeks, who fought the Chinese for Hill 381, west of Inchon, in the January advance towards Seoul: commanded by dramatically moustached Colonel Darboujis, they were attacked three times by 3,000 Chinese, and three times the swarthy men with names like Popadopolous drove them off with bayonets and knives: at dawn the snow was streaked with blood where the enemy had dragged off their dead.

Ask the Turks . . . but then the Turks have never acknowledged the word invincibility if applied to any soldiers – except possibly Turks.

Ask Lieutenant Malcolm Cubiss of the 5th Fusiliers. A studious-looking and bespectacled officer – his thick, horn-rimmed glasses would have excited incredulous comment from any optician in Korea – who personally threw seventy grenades into their midst; ask Fusilier Jim Barker of the same battalion, a six-foot three inch giant from Gateshead, who strode into a section of Chinese armed with a rifle and bayonet and scattered them like chaff.

Ask Second-Lieutenant Mervyn McCord and the men of the Royal Ulster Rifles' 'battle patrol'; ask that experienced and indefatigable trio, 'Yammer' Youds, Billy Reeves, and John Muir.

Ask the 'Princess Pat's', and particularly two bren gunners of 'B' Company, Privates Bill Mitchell and Ken Barwise, who were credited with twenty Chinese each at a range of about ten yards.

Ask any 'Ratty Digger' of the 3rd Royal Australian Regiment: Lieutenant Pat Denness, Private 'Bluey' Smith, Lieutenant Reg Saunders (the Aborigine); ask the Kiwi gunners of 16th Field Regiment of the New Zealand Artillery.

And if the myth of Chinese invincibility is not yet finally exploded, go to Corporal 'Evil-eye' Everleigh, one of the rattiest diggers of them all, whose patrol killed thirty enemy without loss to themselves; New Guinea veteran, and winner of the Military Medal in the October fighting, he reappeared by means still unexplained in the 1st Battalion much later in the campaign. From Everleigh you will get the most picturesque denial of all.

Said Everleigh: 'If them flamin' Chinks are invincible, then my bum's a frying pan.'

Nothing could be much more definitive than that . . .

[8]

April 1951 was a momentous month in the Korean War, not least because it saw the dismissal of General Douglas MacArthur as Supreme Commander of the United Nations forces. He was not only fired from his Korea command, but stripped of his other three hats as well – Supreme Commander for Allied Powers in Japan, Commander-in-Chief Far East, and Commanding General United States Army, Far East. There was no question of MacArthur being 'kicked upstairs', for only the President and God were senior to him; President Harry Truman peremptorily fired him from everything. The resultant uproar in America was at least as noisy as the war itself.

It was no secret that MacArthur violently opposed United States and United Nations policy in Korea. As a military commander – in many opinions one of the world's greatest – he believed that the object of war was victory, and in his book there was no substitute for victory: a war was either won or lost, and this one unquestionably had to be won. In a many-hued publicity campaign he complained long and loudly about the enemy's 'priviliged sanctuary', and declared that 'never before has the patience of man been more sorely tried'.

MacArthur had defeated the North Korean Army; the aim of the United Nations Army, he considered, must be to defeat Red China also. To this end he advocated an intensification of the economic blockade against the China coast, the imposition of a naval blockade, and the employment of Chinese nationalist forces based on Formosa. Using more guarded language, MacArthur also proposed the 'removal of restrictions on air reconnaissance of China's coastal area and of Manchuria'; this, in effect, meant the all-out bombing of Communist air bases, and there was scarcely an American airman in Korea who was not in favour of this – almost every day they witnessed the mortifying spectacle of Manchuria-based and Russian-built MIG fighters rising unchallenged to swoop on American planes with every advantage of height and position.

It was undeniably MacArthur's duty to recommend such actions; but equally it was his duty as a soldier to accept the decision not to implement them. It was here that he made the cardinal error of a soldier: politicians can and always do interfere in army matters, but soldiers, if they want to keep their jobs, must keep clear of politics. MacArthur did not: by judicious, and on occasions unscrupulous, use of the press he attempted to get American and world opinion on his side, and it speaks well for his publicity campaign that he all but succeeded.

At all events, the United States were on his side, if the wildly cheering crowds in Washington on his return were anything to go by; within twenty-four hours of his dismissal, 1,700 telegrams reached the White House protesting against his removal from command; Senator McCarthy, at his vitriolic best, publicly declared that the President was 'a son of a bitch who ought to be impeached'.

President Truman said: 'We do not want to see the conflict in Korea extended. We are trying to prevent a world war – not to start one', and when the uproar had subsided most people in the United States and the non-Communist world agreed with Truman, the United Nations, and Winston Churchill.

And so Douglas MacArthur disappeared from the Korean War scene, but not into obscurity; he addressed both houses of congress in a memorable speech, outlining, explaining, and defending his policies in the Far East. He was placed on the active list for life as a General of the Army. He was succeeded in Supreme Command by General Matthew Ridgway, and command of the 8th Army devolved on Lieutenant-General James Van Fleet.

[9]

Now it was time for the United Nations to take the offensive again: if 'bug-outs' were defeating to morale, General Matthew Ridgway was of the opinion that static warfare was equally so.

The omens seemed good: in April had come a merciful release from the murderous cold; the mornings were clear and bracing; throughout the United States Army morale was on the upgrade, and the American soldiers were beginning to get mad at the 'goddam Chinks' who had kept them away from the comforts of home and Japan for so long; there were many fewer soldiers afflicted with the 'bug-out' mentality, for the winter had weeded out the fainter hearts long since – during the recent Arctic spell a man only needed to remove his boots and socks for a few seconds to become a hospitalized frostbite casualty.

Seoul was taken by the Americans on March 14th and on April 3rd they recrossed the 38th Parallel. During the next three weeks, in a series of limited attacks in the east, the 8th Army pressed forward with persistent caution to secure a line fifty miles north of the stabilized January line on the west. This became known as the 'Kansas Line'.

Ridgway's plan was to push through in the centre of Korea, and to advance so far that the Chinese would have to abandon Seoul. But Ridgway did not want Seoul itself – he regarded it as just one more piece of useless 'real estate', with only prestige value: he aimed to prevent a build-up of any large-scale Chinese offensive, to keep them off balance and ultimately to destroy their armies in detail.

It was a bold plan, and at first it looked like working well: on April 9th the forces of the United Nations south of the 'iron triangle' between Chorwon, Pyongyang, and Kumhwa, started to advance; on April 21st the United States 1st and 9th Corps began to push forward from the line below the triangle, known as 'Utah'.

But Ridgway's plan ran into trouble when the Chinese launched their counter-offensive. Brigadier Tom Brodie's 29 Brigade was deployed on a front of 13,000 yards south of the Imjin River.

It was on this brigade that the Chinese launched the most frenzied assault of the entire Korean War.

PART THREE

The Glorious Glosters – and others

APRIL 1951 – JULY 1952

ELEVEN

THE stage is set for the drama: the river and the surrounding hills.

Firstly, the river: like a sluggish snake, it flows east to west, following a reluctant and exaggerated corkscrew course with bends up to two miles in depth, through a wide and fertile valley, to the sea. At first sight it is a friendly rather than a formidable river, nowhere more than three hundred yards wide and in few places more than twelve feet deep; almost everywhere it is fordable to men and vehicles.

Many rivers in the world have taken their place in military history: the Aisne, the Little Big Horn, the Marne, the Nile, the Po, the Rhine . . . this particular river was the scene of a battle every bit as bloody and hard-fought as these older epics. This river is the Imjin.

Then the hills – a mass of climbing, steeply pointed scenery; they are steep, but not unduly high or precipitous: the tallest is the jagged, 2,000-feet-high Kamak-San. In April the greenery of spring has not yet put in an appearance; the only colour to be seen is the pink of azaleas.

The heroes of the piece are there: to the extreme left the 1st Gloucesters with 'C' Troop of the 170th Independent Mortar Battery of the Royal Artillery; the Gloucesters have not, as yet, been committed to really hard action and are anxious to prove that anything that the 5th Fusiliers and Royal Ulster Rifles can do they can do better. About five miles to the right of the Gloucesters are the Royal Northumberland Fusiliers with 'A' Troop of the Mortar Battery, 45th Field Regiment and the 8th Hussars in close attendance; still farther to the right the Belgian battalion and 'B' Troop of the Mortar Battery.

In reserve are the 1st Royal Ulster Rifles, sadly depleted – they suffered 208 casualties in killed, wounded, and missing in the

January ambush – but still one hundred per cent belligerent.

The 60,000 villains are not yet on the scene – they are about twenty miles the other side of the river – but they are awaiting their cue, and it cannot be long in coming . . .

[2]

It was peaceful enough on the Imjin that April, and the unspeakable winter had given way to a benign and balmy spring:

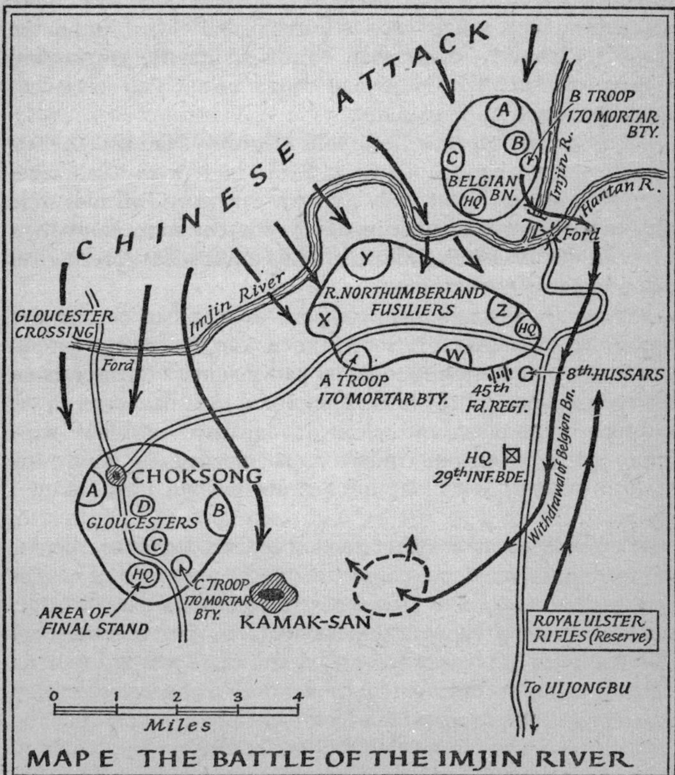

MAP E THE BATTLE OF THE IMJIN RIVER

© CASSELL & CO LTD 1967

the rice paddy now contained not ice but water; the ancient Korean name 'Chosen' – 'The Land of the Morning Calm' – seemed appropriate enough to Korea at this time: in spite of the smell from unburied dead, half-empty tins of food, and assorted ordure, which pervaded the whole Peninsula, life was pleasant enough for the men of 29 Brigade.

There were, of course, the daily and nightly patrols: arduous treks into the twelve miles of no-man's-land on the north side of the river, and the listening posts on the river bank: there could be no let-up in these patrols if Chinese intentions were to be divined. But they were no more exacting and scarcely more dangerous than all-night exercises on Salisbury Plain, and to the reservists infinitely easier than patrols in desert, jungle, and German-held areas of France and Belgium – and that was something for which to be thankful.

These patrols over the Imjin told Brigadier Tom Brodie little enough about Chinese intentions, for Chinese were hardly ever seen. Speculation in the early days was rife in 29 Brigade: were they to continue to advance or stay where they were, ready for a Chinese attack? There appeared to be no immediate prospect of doing either of these things.

Meanwhile, life was more congenial than it had been at any time since their arrival in Korea: men who had been on night patrol slept; men waiting to go on patrol sunned themselves on the reverse slopes of the hills. They stood to at dawn and dusk; weapons were oiled and cleaned; defensive positions were improved with enough industry, as dictated by Company sergeant-majors, to prolong life but not enough to bring on a hernia.

There was no shortage of food, beer, or cigarettes. Cooks, stripped to the waist, peeled potatoes in the open air and sang as they worked. Letters to loved ones at home were long overdue, and they were written with appropriate references to numbers of Chinese killed. Appetites were good and there was virtually no sickness of any sort – even the professional malingerers gave up rehearsing assorted groans which they were saving for the specialists: there was no point in going sick on the banks of the Imjin, and nowhere to go if they did: in any event the medical officers, kindly but cynical men, could spot a malingerer at fifty paces.

The men of 29 marvelled at their luck, and were happy to play along with it. Of course, it was too good to last – something had to happen soon . . .

Every unit had celebrated Christmas Day, 1950; the Scots had had their Hogmanay, and Saint Patrick had had his due at the hands of the Royal Ulster Rifles. But the Royal Northumberland Fusiliers had a very special day in their regimental calendar, and that day was April 23rd – St George's Day: the 5th Fusiliers, wherever they are and whatever the climatic or battle conditions, wear red and whites roses in their caps. They also consume inordinate quantities of food and drink, when available: on the Imjin River they were, but the Fusiliers were to be deprived of them . . .

The Reverend Sam Davies, Chaplain to the Forces (4th Class) of the 1st Gloucesters, travelled astronomical distances, frequently on foot, with mail, cigarettes, and assorted 'goodies', not always obtained by strictly Christian means; his church services were stimulating and uplifting, and regularly attended by men who had 'Atheist' and 'Agnostic' defiantly written on Page One of their pay books. Here was a true man of God . . .

The Royal Northumberland Fusiliers were temporarily without a chaplain, and Colonel Kingsley Foster had invited Sam Davies to conduct and preach at the battalion's St George's Day service on Sunday, April 22nd.

It was a good day for the worship of God, more specifically it was a good day for the St George's Day service, even though it was a day too early. Clearly, St George was all right, thought Davies, as he alighted from his jeep and saw the serried ranks of bare-headed fusiliers; the trouble was the dragon, in the shape of China's wealthy manpower account, twenty miles on the other side of the Imjin and indubitably poised for an annihilating blow. In the meantime – and the Chaplain took refuge in comforting cliché – God was in His Heaven, and all seemed right with the world . . .

All People That On Earth Do Dwell, *Fight the Good Fight*, and perhaps the most stirring and moving hymn of all, *Oh Valiant Hearts*, rose courageously into the blue sky: the scene was one to put war, where it rightly belonged, on the low level of gross indecency. It was not always easy for a priest to promote the Kingdom of Heaven to eight hundred unreflective and hard-swearing soldiers, but on a morning like this the task was made appreciably easier.

Colonel Foster read the lesson, and the words were singularly appropriate to the man – the words of St Matthew: 'For I am a man under authority, having soldiers under me: and I say to this man, Go, and he goeth; and to another, Come, and he cometh . . .' Then the Chaplain told the men something of St George, the early Christian martyr who defied the Romans. After the Service the Commanding Officer proudly read the awards for service in Korea recently conferred on members of the battalion by King George the Sixth: the Distinguished Service Order for Major Reggie Pratt; the Military Cross for Major Charles Mitchell and Lieutenant Malcolm Cubiss; the Military Medal for the Gateshead Giant, Fusilier Jim Barker, who was so overcome by pride and emotion that he reported sick that very afternoon: he was awarded one day's light duty, and much good it did him . . .

Finally to the Commanding Officer's tent for sherry: Kingsley Foster was a man who liked comfort when he could get it, and he had a bottle of a tolerable brand.

Then for Sam Davies it was back to his own parish, the 1st Gloucesters. He set up his altar in a ruined temple in 'A' Company's lines, overlooking the Imjin River, and celebrated Holy Communion. Two subalterns of 'A' Company were among the communicants: Lieutenant Philip Curtis, 24 years of age and recalled from the reserve to fight in Korea, and Lieutenant Terry Waters, recently commissioned from Sandhurst.

> The Body of our Lord Jesus Christ, which was given for thee, preserve thy body and soul unto everlasting life. Take and eat this in remembrance that Christ died for thee, and feed on him in thy heart by faith with thanksgiving . . .
>
> The Blood of our Lord Jesus Christ, which was shed for

thee, preserve thy body and soul unto everlasting life. Drink this in remembrance that Christ's Blood was shed for thee and be thankful . . .

Walking slowly along the line of communicants, speaking the hallowed words and delivering the Blessed Sacrament to these young men all meekly kneeling, it seemed to the Chaplain that a sudden ending to such young lives was unimaginable. But for Philip Curtis, Terry Waters, and a dozen other communicants this service was a Viaticum . . .

[3]

On that tranquil Sunday, in the steep, rugged hills surrounding the battered and deserted village of Choksong, the 1st Gloucesters were ready for the attack which they knew must come.

On the hill feature known as Castle Hill, commanding the long spurs that rose from the southern bank of the river and the dirt road that ran through the cutting along the right flank, was Major Pat Angier's 'A' Company, scene of Chaplain Sam Davies's recent (and, as it turned out, last) Communion Service. Across the road was 'D', guarding the eastern flank of the road that led from the Imjin. 'B' and 'C' Companies, commanded by Majors Denis Harding and Paul Mitchell respectively, were located on hill features, some 2,000 yards south of the river, with their backs to the towering Kamak-San.

There was another hill, and it was later to be known to the world as 'Gloster Hill', but that time was not yet: it still bore the name by which it was known in the Command Post – Hill 234 – and as far as the occupants were concerned, it was exclusively their property.

They had, undeniably, made it so: the occupants of Hill 235 were the Pioneer Platoon of the 1st Gloucesters, and it was the most exclusive domain in the battalion, if not in the whole of 29 Brigade. For the pioneer platoon, with the exception of half a dozen young National Servicemen, were the men who knew it all – a seasoned, well-knit body of 'old hands', who had been yanked out of comfortable civilian life back to the Army, and were determined to sweat out their compulsory time in the maximum of comfort.

Nowhere in the Gloucesters' position were the trenches more deeply dug, more skilfully riveted, more impervious to shellfire or any other sort of fire; nowhere were rations more plentiful, or the issue of 'goodies' on a more sumptuous scale; nowhere were there more cigarettes – every Gloster had done well out of the compulsory sacking of Pyongyang, but the pioneer platoon were in a special category of acquisitive opulence, and Private Charlie Edkins was the proud owner of something like 5,000 – Chesterfields, Lucky Strikes, Camels, Old Gold, and Philip Morris: just name the brand, and Charlie Edkins had them. On Hill 235 the 'brew-ups' were the most frequent, and the tea the strongest; the card schools the fiercest; the language the most robust; the letters home the fewest; the shaving the closest; the conduct sheets the most comprehensive.

But quickly perish the thought that the pioneer platoon got away with anything, for their platoon commander, Captain 'Spike' Pike, was the most knowledgeable old sweat of them all: in 1936, aged 16, he had joined the army as an apprentice instrument mechanic; by 1938 he was a gunner on 'man's service' in the Royal Artillery, and in 1942 was storming ashore as a sergeant in Lord Lovat's 4 Commando. However much the men of the pioneer platoon knew about the subtler nuances of old-fashioned soldiering, 'Spike' Pike knew a damned sight more.

Hill 235 was 'Pike's Hill' – a cherished piece of 'real estate' – and 'Spike' Pike had made it so. It was with some regret that 'Spike' viewed his imminent departure from this little kingdom: his time was up in Korea, and on April 22nd he was waiting for the posting order which would take him out of Korea, via the fleshpots of Hong Kong, to a comfortable billet in the Royal Army Ordnance Corps at home. Already with him on Hill 235 was his successor, Second-Lieutenant Alan Blundell: 'Spike' had handed over the explosives, the picks and shovels, and the bewildering array of stores which go to make up a pioneer platoon; the platoon had looked at Blundell with appreciative condescension – only a nipper, they said; we'll soon have him eating out of our hands. 'Spike' had indicated the men to Blundell with the terse injunction, 'They'll be all yours soon, and God help you'. For all that, 'Spike' had a fierce affection for these men, and every one of them would cheerfully have let the platoon commander walk on their faces if it had afforded him any satisfaction.

Private Jack Biddle was universally known as 'Uncle' to the young soldiers, because he was constantly giving them advice of a spurious nature – advice such as a dissolute old *roué* gives to an impressionable young nephew on a weekend in Paris: if the youngsters had followed 'Uncle's' advice, they would have qualified for immediate admission to a detention barracks – and on this particular aspect of military life 'Uncle' was ready with advice also. He was even known as 'Uncle' to the Korean porters. Replete, ribald and supremely confident in their ability to give of their best in battle, the pioneer platoon watched and waited . . .

Scattered about the hills were Captain Theo Littlewood's machine gunners and Captain Graham Lutyens-Humphrey's mortars: belts of ammunition and bombs were ready. So, too, were the lay and spiritual healers: Captain Robert Hickey, the Medical Officer, and Sam Davies, the Chaplain.

In the command post were Captain Anthony Farrar-Hockley, the Adjutant, and Lieutenant Henry Cabral, the Intelligence Officer. And in this command post, at all hours of the day, there was a certain smell – or perhaps aroma is a kinder word.

The two principal smells of the Korean War were the dead and the dung. But in this command post was a homely and comforting smell – the smell of a good quality English navy cut, burning slowly and evenly in a well-seasoned briar pipe; a smell as English as the fragrance emanating from a fish and chip shop.

The third occupant of the command post was the owner of the briar pipe. This is the Commanding Officer of the 1st Gloucesters, Lieutenant-Colonel James Power Carne, universally known as 'Fred', and the greatest Gloster of them all . . .

In a tent in the Battalion Headquarters area a party was in progress. Outside this tent was a sign, like *Mon Repos* or *Bella Vista* outside a suburban villa; but this sign said simply, 'RSM', and only a very chosen few stepped inside. According to the awed description of one private who went in to deliver a message, it resembled a suite in the Grand in Broad Street, Bristol (not that this soldier had ever seen one): there was, he declared to a gullible audience, a four-poster bed, with a deep sofa, a built-in cocktail bar, and a chamber-pot with the regimental badge on it.

Most of the Gloucesters discounted this Bristolian extravagance, but the tent was undoubtedly well-appointed within: RSM Jack Hobbs, after a quarter of a century of service (five of them spent in a prisoner-of-war camp in Germany) was a firm believer in the adage 'rank hath its privilege'; he liked his comforts and considered himself entitled to them, and any griping soldier who didn't like it could come and see him about it – personally . . .

Besides being well-appointed, it was also well-victualled on the night of April 21st, 1951. For this night signalled an occasion – the type of occasion which is frequent in Sergeants' Messes in every fortress, at home or abroad, where the Union Jack flutters.

An august personage, Orderly Room Quartermaster Sergeant Taffy Evans, no less, had come to the end of his military service. He had done his 'pontoon', and 'one for the King', and to the layman this period of time amounts to twenty-two years; like Spike Pike, he was due for England, home and beauty, but not for the Royal Army Ordnance Corps: for Taffy Evans, it was to be 'Civvy Street'.

The ORQMS is a very essential key man in any battalion, and without him the Adjutant would be virtually crippled; he knows that Private A will become a lance-corporal if he stays out of trouble for the next six months; he knows the exact date of Private B's last red ink entry for being drunk; he knows that Corporal C's marriage allowance is not being paid to Mrs C, but to a cinema usherette in Weston-super-Mare; he knows that Sergeant D will be eligible for promotion to Colour Sergeant when he has surmounted that last daunting hurdle, the 1st Class Certificate of Education. Not to put too fine a point on it, he knows everything.

Also present, half envying and half commiserating with the ORQMS, were other weighty Battalion Headquarters personnel and the combined total of their service came to nearly one hundred years: CSMI 'Muscles' Strong, the PT instructor and Commanding Officer's personal bodyguard; Drum-Major ('Drummie') Philip Buss, whose skilful handling of the Beating of Retreat ceremony in Colchester had attracted the favourable attention of the Mayor himself; Sergeant Jim Smyth, the signal sergeant; Sergeant Bill Peglar, Chief of the Gestapo, otherwise the Provost Sergeant.

But ORQMS Evans little knew that there was to be no discharge for him – not just yet . . .

The most lonely and worried man in 29 Brigade on the afternoon of April 22nd was Brigadier Tom Brodie: lonely, because once he has committed his forces, a commander is, of necessity, a solitary man with virtually nothing to do; he carries an enormous burden of responsibility and carries it alone. Brodie visited all his units, and found them in good heart: the prevalent feeling in the Gloucesters, the Northumberland Fusiliers and the Belgian battalion was 'let 'em all come, and the more the merrier'.

Heartening as this was, Tom Brodie was still worried, although no one in his Headquarters would have guessed it. Nor would the press correspondents: on the evening of April 21st Brodie had visited the press camp, a cheerful and extrovert establishment where liquor was plentiful and rumour perpetually rife. The pressmen bombarded Brodie with questions, and the most universally repeated one was: 'When's the Chinese offensive coming?' – thus implying some clairvoyant quality that the Brigadier wished he possessed.

'Your guess is as good as mine,' said Brodie, and applied himself with enthusiasm to a large whisky and soda.

Nevertheless, Brodie *was* worried, and he had plenty to worry him: his force was woefully thin on the ground, and his pleas for reinforcements had been met by the American higher command with sympathy but total lack of action – 'Brigadier, you've got to understand that it isn't that we don't *want* to give you more guys, we just don't have them. But we know you're gonna hold': Tom Brodie, like Aubrey Coad in an earlier campaign, found this naïve trust in British indestructibility gratifying, but it did not alter the facts. And the facts were that trying to hold the approaches to Seoul with 29 Brigade was about as feasible as blocking the approach to London by stringing soldiers across Ascot racecourse.

To a soldier, Sandhurst and Staff College trained, and with a quarter of a century of service, much of it active, the gaps between battalions seemed terrifyingly large: to Brodie's experienced eye, it looked as if a complete Chinese division could infiltrate with consummate ease between the Gloucesters and the 5th Fusiliers and another one between the Fusiliers and the Belgians.

There were no reinforcements: there were only Gloucesters, 5th Fusiliers, Ulster Riflemen, and Belgians with rifles, bayonets, bren guns, and guts. And with these Brodie had to be content...

[4]

For nearly three weeks now there had been this brooding and sinister activity on the far bank of the Imjin. But on that sleepy afternoon of April 22nd – not an afternoon for war, because men had been to church, lunched well, and were sinking into an agreeable torpor of sleep – Chinese were seen: here a group of ten men, patrolling casually and pausing to sweep the opposite river bank with field-glasses; there a group of twenty, moving rather faster, and then what appeared to be at least a hundred – sizing up the situation, appraising United Nations strength on the other side. For a force allegedly bent on 'Imperialist aggression', it must have seemed pathetically small to the Chinese Commissars.

Lying on their stomachs watching this renewed Chinese activity were three officers of the Gloucesters: the Adjutant, Captain Tony Farrar-Hockley; the Intelligence Officer, Lieutenant Henry Cabral; and a thin column of blue smoke told that the other watcher was the Commanding Officer, 'Fred' Carne. And as they watched, with quickening pulse rates, these three men knew that the period of peace was nearing its end.

Said the Colonel: 'This looks like the real thing. Of course, it may only be a feint or some sort of deception force, but I doubt it. In any case, it's high time the mortars did a job of work.'

Captain Graham Lutyens-Humphrey needed no second bidding: there was a series of flashes, and then puffs of black smoke as the bombs landed among the carelessly wandering Chinese.

Colonel Carne gazed through his glasses for a few minutes, his pipe drawing unhurriedly. Then he gave his orders to Farrar-Hockley, and they sounded the knell of peaceful sleep for that night: there would be a 50 per cent 'stand-to' and, in anticipation of an attempted enemy crossing of the river, an ambush patrol on the river bank. With a tactical sense, which sometimes bordered on the uncanny, Carne was reading the mind of the Chinese commander: he would, Carne deduced, send a party across the river that very night, and the Gloucesters were going to be ready for them.

The honour of striking the first Gloucester blow – the first of many – fell to No 7 Platoon of 'C' Ccmpany, commanded by Lieutenant Guy Temple, universally known as 'Guido', because his dark hair and swarthy complexion gave him an oddly Italian appearance. Just after darkness fell Temple positioned his sixteen men on the river bank and waited.

Every infantry soldier dreams of such a situation: a well-sited force of trained and determined men, a generous allocation of automatic weapons, a full moon: Temple had all of these things, and the river was bathed in a translucent glow.

The bren guns and sten guns were cocked; safety catches of rifles were forward; fingers were wrapped round the pins of grenades. Uppermost in every man's mind was the thought that if the Chinks were fools enough to cross the river at this point, then they deserved all that was coming to them.

Mindful of the no talking order; mindful, too, of the dire consequences threatened by Temple if any man opened fire before getting the order, the men of No 7 Platoon, with unbearably itching trigger fingers, watched and waited.

Then suddenly there was a scuffle and a splash: four Chinese slipped into the water and started to cross, then three more. Soon the river seemed to be full of Chinese, all clearly illuminated by the moon, all sitting targets, all blissfully unaware of the Gloucesters who were waiting to kill them.

If Temple's men did not actually wait until they could see the whites of their adversaries' eyes, they did the next best thing: at a range of approximately ten yards they opened up with everything they had, and the resultant slaughter was a triumph for fire discipline.

The Chinese, mercilessly exposed as they were, never had a chance: some yelling at the tops of their voices and firing wildly as they stumbled through the water, came almost to the very muzzles of No 7 Platoon's guns: others turned about with only thoughts of flight to guide them, and were ignominiously shot in the back; many more, dead and dying, were swept downstream in water now liberally stained with blood. It was success, complete and absolute for the ambush party; in the space of thirty seconds they had dispatched as many Chinese.

'Nice job,' commented Guy Temple; 'but they'll be back. Check your weapons and keep your eyes skinned.'

And they came back in ever increasing numbers: dozens became scores, and quickly multiplied into hundreds. Two, three, four times the Chinese attempted to cross, and four times they ran into the same deadly and controlled fire: riflemen picked off the struggling men in the water as fast as they could work their bolts and machine gunners sprayed the water, by now plentifully reddened with blood. Obscenely bathed in the bright moonlight, the river became a dark shambles of shrieking, struggling, drowning Chinese. Not until the last yelling voice was silent and the last desperately clawing hand still, did Guy Temple give the order 'cease firing'.

There was not a man in the ambush party who had not killed at least three Chinese. Private Len Allen, a bren gunner who had performed with indifferent success on the range at Colchester and had severely shaken a general on the dockside before sailing by announcing that he had no quarrel with the Chinese, counted ten corpses in front of his position; said an incredulous reservist, who had fought the Japanese in Burma and was accustomed to furious onslaughts of enemy infantry: 'The b— must be breeding over there.'

But now bren guns were down to less than one full magazine each and riflemen had a bare five rounds apiece. They had beaten off four attempts to cross, but Temple knew that the next rush would swamp them.

'Our job's done here,' said Temple to his men; 'it's time to go back.' Elated and unhurt, the men of No 7 Platoon slipped noiselessly away from the river bed and returned to 'C' Company's position. Their job may have been finished there, but there would be plenty more for them to do soon . . .

[5]

On Christmas Day 1953, a grandmother and a little girl – a pigtailed, gap-toothed, wide-eyed mite aged four years – sat together in the modest sitting room of a modest terraced house in Devonport. To Philippa Curtis it seemed a funny sort of Christmas, because there were no men there: no Father Christmas (although he had been the night before, as a bulging stocking and a generously arrayed tree testified), and no Daddy. Perhaps Father Christmas would put in a personal appearance later in the evening for some cake; perhaps, with a bit of luck, Daddy would be home for Christmas next year . . . clearly, in this family Daddy was an important person, for Philippa's mother had died while giving her birth. And that only left Granny . . . Granny was good, of course – ever ready with food, bedtime stories, the requisite amount of spoiling, the bag of sweets at exactly the right moment, and the warm kiss before going to sleep every night, but it wasn't quite the same thing.

Two frames on the crowded mantelpiece told that Philippa's father was an important person indeed: in one frame was a photograph of a serious-looking young man in army uniform, a not very impressive moustache showing under a cap sitting very straight on the head, Gloucester style. In the other frame was a medal – the dark red ribbon and the simple bronze cross, with the inscription 'For Valour': this was the Victoria Cross, posthumously awarded to Lieutenant Philip Curtis of 'A' Company, 1st Battalion the Gloucestershire Regiment, on the night April 22nd–23rd, 1951, for 'magnificent conduct throughout this bitter battle'.

It was a bitter battle indeed, and Philip Curtis was in the thick of it right from the start; although Guy Temple's ambush party had inflicted very considerable casualties on the enemy at 'Gloster Crossing', the Chinese corpses washed downstream by the swirling river were, so to speak, on the Chinese expense account: China's wealthy manpower account had not, as yet, been seriously tapped.

The first frenzied assault fell on Major Pat Angier's 'A' Company and the Chinese, as always, came head-on, heedless of casualties: there was no time for fire discipline now – bren guns were fired until the barrels became red hot, rifles until they were too hot to hold; Vickers guns pumped belt after belt of ammunition into the yelling hordes until the cooling jackets of the guns boiled over and the guns seized up. The battalion mortars, 170th Independent Mortar Battery and the guns of 45th Field Regiment poured bombs and shells into the closely packed masses assembling to reinforce the assault.

No accurate count of enemy casualties could be made, but for every Chinese killed or wounded there were five to take his place.

The battle was barely half an hour old before the first grave reports came in to Colonel Carne at battalion headquarters, and from then on they came in almost continuously: a platoon overrun; a position lost, retaken by counter-attack with the bayonet, and lost again; a platoon reduced to six men; an officer killed; two more officers wounded; machine guns knocked out one by one; more enemy massing; appeals for reinforcements and artillery fire; a platoon virtually wiped out; another officer killed . . . there was no sleep for Carne or for anyone else that first night, and for hardly a second of time was the slow-burning pipe – that symbol of imperturbability – out of the Colonel's mouth. He never raised his voice or swore; his calm voice giving orders to his company commanders never once wavered or shook; he was calm and strong at all times, and the other officers and men in the headquarters dug-out stole an occasional look at him, and so became calm and strong in their turn.

Very early in the fighting Second-Lieutenant John Maycock was killed and his platoon reduced to six unwounded men. Before they could be completely wiped out, Pat Angier withdrew them to make very welcome additions to Philip Curtis's platoon, also considerably reduced in numbers: the time was not far off when a platoon consisting of six whole men was to become almost commonplace.

As Angier's men continued to hurl back one assault after another, Major Denis Harding's 'B' Company was also being subjected to attack.

But the first attack on 'B' Company was carried out with probing caution; initially it had only nuisance value, and was summarily and bloodily repulsed.

Earlier that night Private Albert Robson, a bren gunner in Lieutenant Geoffrey Costello's platoon, had failed to distinguish himself: for two hours he had been straining his eyes, staring out into the inhospitable blackness in front of his section's position. Just ahead of him he heard the crackling of a twig, then another: this could mean only one thing – there were Chinese out there and it was up to Albert Robson to dispatch them immediately, if not sooner. He cocked the gun and fired a long and uncontrolled burst into the middle distance. He was in the act of placing another magazine on the gun, when the large figure of his section commander, Corporal Sam Juggins, loomed out of the darkness.

'And where the hell do you think you are?' inquired the Corporal with enormous menace. 'Bisley?'

'Thought I saw some gooks,' said Robson.

'You never saw nothing,' said Juggins unkindly. 'All you did was let the gooks know we're windy. Next time I hear that gun of yours I want to see a stiff in front of it. Got it?'

Half an hour later, Private Robson made no mistake. He withheld his fire until the enemy were virtually on top of him and killed thirteen Chinese.

'That's more like it,' said Corporal Juggins.

Similar probing attacks were made on 'B' Company throughout the night, but all were beaten off with almost contemptuous ease: a party of Chinese approached Corporal Bob Crisp's section and ran into a veritable hail of hand grenades; Private Ron Maskell, winner of many a shooting trophy in peacetime, scored a 'possible' – three shots, three dead Chinese; Arthur Peal, a keen and irrepressible Second-Lieutenant, just six months out of Sandhurst, jubilantly counted thirty bodies in front of No 6 Platoon's position.

As dawn broke on St George's Day, 'B' Company were in fine fettle: at no loss to themselves, they had killed at least one hundred Chinese, and the reservists were telling the young soldiers that this war was a picnic compared to Caen, Cassino, Kohima, and other sanguinary Second World War battlegrounds. Morale was sky-high, and there was hardly a reservist who would not have signed on for another seven years there and then . . .

There was nothing wrong with 'A' Company's morale, but in every other respect they were in evil case: during the night the fighting had been vicious and continuous, and although Angier's men had inflicted astronomical casualties on the Chinese, they had suffered grievously themselves: half their number were dead or wounded; the remainder, exhausted, red-eyed but indomitable, crouched in their weapon pits, waiting for the new attack which they knew would come. And all the time they were lashed by machine-gun fire.

This fire came from the hill feature known as 'Castle Site', vacated during the night by the half dozen survivors of John Maycock's platoon. 'Castle Site' was the highest point of 'A' Company's position, and there the Chinese had set up machine-gun posts from which they could spray the defenders with insulting ease.

Casualties continued to mount steadily: here was a man, dead, shot neatly between the eyes; here two more – they had shared a burst, and now had one serviceable leg between them; here a man with six machine-gun bullets in the stomach who, with the best of medical care instantly available, could not hope to live for more than an hour.

Tending the wounded with selfless devotion was Corporal Jack Papworth, known in 'A' Company as 'The Quack': cutting away shirts and trouser legs, staunching bleeding, applying field dressings, making serious inroads into his personal cigarette supply acquired in Pyongyang.

It was obvious to Pat Angier that so long as the Chinese remained on Castle Site, his company would suffer even more casualties, if not be wiped out altogether: the machine guns dominated the platoon positions of Philip Curtis and Terry Waters; unless Castle Site was retaken, and quickly, the Chinese would make a further annihilating advance which would endanger, not only 'A' Company, but the whole of the Gloucesters' position.

Angier's orders to Curtis were brief and to the point: 'You've got to shift 'em off it, Phil.'

Curtis said 'Right', and gathered his men together for an assault which, viewed by any standards, seemed well-nigh suicidal.

To achieve an effective frontal counter-attack against Castle Site Philip Curtis needed artillery support on a generous scale, preferably a pulverizing airstrike and eager, fresh and well-fed infantry soldiers to finish the job. Curtis had none of these things: he had twenty men with painfully rumbling stomachs – their last square meal had been at midday on the 22nd – and all of them desperately tired from a night of continuous fighting.

At the head of his tiny force, Curtis advanced across open ground: there was not a stick of cover anywhere, and they were raked with accurate machine-gun fire: three men fell dead after they had advanced twenty yards; four more were wounded sixty seconds later. Then Curtis himself was hit, in the right arm and left side; all impetus seemed to have gone out of the counter-attack before it had properly started.

The ubiquitous Corporal Papworth was quickly on hand to deal with the casualties and Curtis, already bleeding profusely, indicated that they should be dealt with first. Then he looked towards the enemy position; it seemed that the intrusive machine gun was now silent, but this, Curtis well knew, was only temporary: as soon as they advanced again, it would open up on them and comprehensively mow them down. Nevertheless, he had his orders from Major Angier: 'The Castle Site will be retaken,' and as long as there was breath in his body, six rounds in his pistol and a grenade in his hand Philip Curtis was going to take it . . .

He shook off restraining hands, as Corporal Papworth, medical haversack at the ready, came towards him. Then Curtis went forward alone: once he fell down, but got to his feet again and continued his lone charge, staggering like a drunken man, leaving a thin trail of blood behind him. Philip Curtis was dying fast, but he had his orders: 'Shift 'em off the Castle Site . . .'

Still miraculously alive, he came to within twenty yards of the machine-gun position; drawing on some incredible reserve of strength, he staggered forward a few more yards and hurled a grenade – a second before the grenade exploded, demolishing the gun and killing the crew, the machine gun emptied its last burst into Philip Curtis's already sagging body.

And that is the story of the photograph and the medal in a modest terraced house in Devonport.

'A' Company of the 1st Gloucesters had become all but a spent force: Lieutenants John Maycock and Philip Curtis were dead; Lieutenant Terry Waters had been badly wounded in the head, and by Corporal Papworth's attentive reckoning would soon be dead too.*

When Major Pat Angier telephoned Battalion Headquarters for orders, Tony Farrar-Hockley had recorded that, for the first time since the battle began, he detected a faint tremor in Fred Carne's hand as he poised a match over the inevitable pipe. For Carne had to give an order which virtually condemned some thirty men to wounds, mutilation, and death.

Not that Angier complained about this order; indeed, he was not greatly surprised by it. He was merely a company commander and a professional soldier reporting a situation of extreme gravity as he saw it himself. On the wireless to Farrar-Hockley, Angier said calmly and regretfully: 'I'm afraid we've lost Castle Site. I want to know whether I am to stay here indefinitely or not. If I am to stay, I must be reinforced as my numbers are getting very low.'

Farrar-Hockley called the CO to the set, and Carne gave an order, and it was this order which explained the sudden clenching of his teeth on his pipe and the just detectable tremor in his hand as he re-lit it: 'You will stay there at all costs until further notice – *at all costs*.'

Anthony Farrar-Hockley marvelled at this Commanding Officer of his, who was seeing his battalion – the battalion he had

* Lieutenant Terry Waters died, but not until very much later. He was taken prisoner, and subsequently awarded a posthumous George Cross for extraordinary heroism in captivity.

served, on and off, for a quarter of a century – shot to pieces all round him. Pat Angier's last words were: 'Don't worry about us; we'll be all right.'

He was killed a quarter of an hour later.

'D' Company, commanded by Captain Mike Harvey, were in scarcely better shape: Mike Harvey, bespectacled, untidily moustached and a devotee of the art of Judo, had been elevated to the command of the company through the strange fortunes and permutations of war: the regular company commander, Major 'Lakri' Wood, was on leave in Japan; on hearing of the Chinese offensive Wood had bombarded every airfield with clamorous demands for air transport to Korea, but there was none available.

Like 'A' Company, 'D' had been incessantly mortared and machine-gunned; they, too, had been subjected to ceaseless attacks by overwhelmingly large numbers of Chinese, but had resolutely stood their ground. But for them also the sands were fast running out; they were gradually being swamped by a furious tide of Chinese.

'B' and 'C' Companies had not, as yet, been in hard action: 'B', indeed, were still indisputably 'kings of the castle', and the order to withdraw came as a mortifying surprise. But Colonel Carne well knew that every minute was seeing a new widening between the fighting groups, and this could eventually only lead to one outcome – the Chinese would isolate them in turn and destroy them one by one.

At about 8.30 PM on the evening of April 23rd, the men of 'A' and 'D' Company started to file into the area of Gloster Hill: their faces were grey with exhaustion, and they staggered into the Battalion Headquarters area like sleepwalkers. All were heavily laden with arms and ammunition; all were in the uttermost extremes of exhaustion, for there had been no sleep for anyone the night before.

Watching the long files as they straggled in, the Colonel felt a great surge of pride: he could see men he had served with before the war; young soldiers to whom he had awarded fourteen days' CB in Colchester; reservists who had come to him with tales of domestic woe of startling complexity – he could even see one who had requested a personal interview with the War Minister.

Came the wounded – the stretcher cases, borne by sweating, stumbling men; the walking wounded with arms in slings and hastily bandaged heads; a young soldier, blinded by the flash of a grenade, walking falteringly with his hands on the shoulders of the man in front.

The battalion's resident healers were ready for them: the Medical Officer, Captain Bob Hickey, went to work at once and his hands were soon red with blood as he dealt with one casualty after another; so, too, were the hands of his helpers – 'Knocker' Brisland, the medical sergeant, Sergeant Baxter, the attached RAMC sergeant; Private Ted Bounden, in the Medical Officer's stretcher-fitted jeep, made frequent forays up the bullet-swept road in search of more wounded; Sam Davies, the Chaplain showed himself adept at ministering physical as well as spiritual comfort.

And, of course, there was Corporal Watkins, the cook: Taffy Watkins had been undaunted on Christmas Day, when all the turkeys froze, and he was undaunted now; there were no turkeys, only sausages, bacon, bread, and tea culled from a fast-diminishing store. In the past Watkins had heard the hoary old chestnut applied to himself: 'Who called the cook a bastard? Who called the bastard a cook?' But now the men of 'A' and 'D' looked upon him, as his busy and seemingly uncoordinated hands summoned up still more 'bangers', as a *cordon bleu* chef, if not a ministering saint.

Now it was the turn of Major Denis Harding's 'B' and Major Paul Mitchell's 'C' Companies. The treatment meted out to them was the same as that to 'A' and 'D': they were raked by close-range machine-gun fire and ceaselessly attacked by waves of screaming Chinese, wildly firing burp guns, machine guns, and rifles and hurling grenades.

In the Gloucesters' command post Colonel Carne, Captains Anthony Farrar-Hockley and Reeve-Tucker, and Lieutenant Henry Cabral drank a hasty brew of coffee.

Carne, puffing composedly at his pipe, came to a decision. 'The time has come,' Tony, he said to his Adjutant, 'to increase the membership of the little club we've got here. We're barely two companies strong, and weak ones at that. We're all going to be one big family: I'm pulling everyone into one perimeter. 'C' Company will start coming back in ten minutes, and 'B' will join us at first light.'

During a brief lull in attacks Paul Mitchell seized an opportunity to extricate 'C' Company. But there were to be no more lulls for 'B'.

Between 11 PM on the 23rd and dawn on the 24th, Denis Harding's men were subjected to not fewer than seven all-out attacks, and the forward sections were overwhelmed by sheer weight of numbers. By dawn ammunition was practically exhausted, and when the hand grenades ran out men hurled beer bottles at the engulfing waves of Chinese. There was hand-to-hand fighting with bayonets, rifle butts, entrenching tools, boots, and bare fists.

When 'B' Company of the 1st Gloucesters joined the rest of the battalion early in the morning of April 24th it consisted of Major Denis Harding, Company Sergeant-Major John Morton and fifteen men. They brought out all their weapons, but an ammunition check yielded just four rounds in Major Harding's pistol. 'B' Company was combined with 'C', and in numbers they amounted to one weak platoon.

The 1st Gloucesters' line had shrunk from four miles to 600 yards, but at no point had it been breached. Now they were all together: rifle companies, signallers, pioneers, drivers, regimental police, cooks, medical orderlies, mortarmen, machine gunners. As the work of redisposing the battalion – now approximately three hundred whole men strong – went on, Colonel Carne said to Farrar-Hockley: 'It seems to me that you and I are going to find a job for ourselves as riflemen before very much longer.'

Colonel Carne recalls one remark made by a member of the mortar platoon as the most heartening he heard during the whole campaign. On hearing that the battalion was to make a last stand, all together on the same hill, the mortarman said in a rich West country burr: 'Us'll be all right, zurr – t'will be laike the Rock of Gibraltar up yere.'

But one of 'Spike' Pike's pioneers took a less exalted view. This was *their* hill – comfortable, exclusive, fortified with loving care. Now 'A' Company, bits of 'C' Company, machine gunners and odds and sods of every description were crowding on to it, scrambling all over the shop like a lot of old whores at a christening. This was nothing less than trespassing; worse, it would cause acute overcrowding.

'What's the game, then?' demanded the irate pioneer. 'This is *our* hill.'

'Not any more it isn't,' corrected Company Sergeant-Major Mike Gallagher of 'A' Company. 'Now it's *Gloster Hill*, and don't you forget it . . .'

[6]

Not only the Gloucesters were being hard pressed: St George's Day was only a very few hours old when the Belgian Battalion and the Royal Northumberland Fusiliers were enveloped by two separate 'human seas'.

'Gallant little Belgium' – the phrase was first coined in 1914, when de Witte's tiny and hopelessly outnumbered army held up von Kluck's 'hordes of fiends' before Louvain. It was not easy to gauge Belgium's interest in a Korean conflict, but the phrase still held good: on the Imjin River the Belgian's fought as if the Chinese were battering at the gates of Brussels itself.

At first the Belgian battalion had been a source of derisive amusement to the more conventional British soldiers: in the words of Hugh Hamill of the Royal Ulster Rifles, they were something of a 'Beau Geste' force; they were not over keen on shaving, and Brigadier Brodie has recorded guardedly that their sanitary arrangements sometimes bordered on the primitive. But they knew how to fight, and beat off Chinese assaults with the fortitude which is a national characteristic.

In common with the rest of 29 Brigade, the 1st Royal Northumberland Fusiliers initially regarded their sailing to Korea as a prelude to perfunctory occupation duties in reasonably civilized surroundings; consequently a cornucopia of fusilier officers' plenty, consisting of dinner jackets, fishing rods, shot-guns, and (so it was whispered) polo sticks accompanied the battalion baggage at rear echelon. It all stayed there.

On the night of April 24th, as the beleaguered Gloucesters prepared for their historic last stand, a furious attack fell on the Northumberland Fusiliers, and more specifically on 'Z' Company. To keep the Chinese off the Fusiliers' position on a knife-edged slope to the right of the Gloucesters required very exceptional soldiers: 'Z' Company had them.

Their commander was Major John Winn. By 1951 he had seen more active service than the majority of soldiers experience in a thirty-year career. During the first ten years of his service John Winn had amassed a list of wounds and injuries which the average man would expect to sustain if he were to be run over by an express train: as a young subaltern in Egypt, in 1936, he fell out of the dickey-seat of a fast-travelling car and sustained a broken back, a fractured skull and the breaking of three vertebrae and four ribs; in North Africa a German shell in his immediate vicinity whipped six pieces of jagged metal into different parts of his body; in the closing stages of the Italian campaign his right arm was shattered by a burst of machine-gun fire. In 5th Fusilier circles it was surmised that the doctor who had passed him fit for service in Korea must have been blind or drunk or both, but John Winn would not have cared what they thought – if the Fusiliers were going to war again, then he was undoubtedly going with them.

The year 1949 found John Winn, as a comparatively elderly captain, in Singapore with no immediate promotion prospects: he occupied a somewhat nebulous staff appointment but coveted no comfortable staff job; he was first and foremost a regimental soldier, and by his uncomplicated reasoning any soldiering done away from the 5th Fusiliers was time wasted and time which weighed heavily upon his hands.

In 1950 John Winn got all that he desired of life: command of a rifle company of his own battalion in war.

Fusilier Ronald Crooks had so far only been exceptional in

that he had been a constant source of trouble to his officers and NCOs ever since he joined the battalion in Bury St Edmunds. He was not just difficult – he was frankly impossible.

He was frequently insubordinate; he was not averse to letting people know that he had been wounded in Italy and had won the Military Medal for some crazy impossibility with a rifle and bayonet; he informed Major Winn, in the most explicit terms, that he was a Durham light infantryman and *not* a Northumberland fusilier, and that the sooner the War Office rectified such a glaring error the better it would be for the British Army in general and Ronald Crooks in particular. To the fury of his NCOs, he insisted on referring to himself with thick irony as *Private* Crooks and not *Fusilier* Crooks; this did not go down well with John Winn . . .

Everyone hoped that he would improve once they got to Korea – indeed, John Winn was beginning to hope that the Durham Light Infantry would claim him back at an early date as a matter of urgent military expediency – but this was not to be: Crooks objected to all the less glamorous military chores – peeling potatoes, he declared forcibly, was a job for a f— kitchen-maid but not for a fighting soldier. It was the same with digging: one morning Major Winn came upon the unedifying spectacle of Crooks, leaning on a spade and smoking a cigarette, while his section corporal slaved on improvements to a trench. It goes without saying that he was a constant thorn in the side of Company Sergeant-Major Edward Radcliffe.

There were many senior officers, warrant officers, and sergeants like Radcliffe at the end of the war. They were, in truth, something of a lost legion – regular professional soldiers who, through no fault of their own, had come through a global conflict unscathed and without hearing a shot fired in anger.

Waiting for the Chinese, Radcliffe, a not over-imaginative man attempted to analyse his feelings and came to the conclusion that he was more apprehensive than frightened; his apprehension was not governed by fear of death or mutilation, but rather by fear of being afraid. Perhaps when the Chinese attacked, when they were running towards him intent on killing, he would be frozen into immobility or he might turn and run in panic. And, of course, as he was the Company Sergeant-Major, the eyes of the fusiliers would be upon him: the young soldiers would look to him for an example; the old hands with battle experience would eye him with speculative suspicion.

But CSM Radcliffe need not have worried . . .

There was Sergeant Nobby Clarke, a short stocky NCO, and the possessor of fierce black moustaches: Clarke had been elevated to the status of platoon commander because his officer, Second-Lieutenant Paddy Baxter, had been whisked home on sudden compassionate leave. Clarke had a domestic problem which was almost commonplace among the 'old men' of 29 Brigade: he had become the father of a baby daughter at almost the precise moment that the 5th Fusiliers disembarked at Pusan – a daughter whom he had never seen and never would see, for Nobby Clarke was to die in the furious attack launched on 'Z' Company.

There was Second-Lieutenant William Shepherd, and he was like many another young subaltern just out of Sandhurst – inarticulately overflowing with pride at the brand new solitary pip on his shoulder, striving desperately for the respect and affection of his men, and hoping nervously to hear shots fired in anger. He was to achieve all these things.

There were Fusiliers Jack Tinsley and Tom Topliss, the company commander's batman and signaller respectively: experienced soldiers both, they knew that their place was by Major Winn's side at all times and were perfectly satisfied with such an arrangement, as was Major Winn.

Something was undoubtedly cooking on the brigade front, and every Northumberland Fusilier fervently hoped that it was turkey: the 5th Fusiliers traditionally celebrate St George's Day in the same manner as Americans commemorate Thanksgiving Day – with inordinate consumption of roast turkey and liquor.

But clearly something had gone wrong with this particular St George's Day. When the first attacks were launched on the Gloucesters at midnight on April 22nd, Colonel Foster had ordered the evacuation of turkeys and alcohol from forward company positions: the possible overrunning and capture of Royal

Northumberland Fusiliers' positions was a normal hazard of war, but Chinese possession of turkeys and liquor which rightly belonged to the 5th Fusiliers was not to be countenanced in any circumstances. Some sort of compromise was arrived at; every man was issued with an outsize turkey sandwich, with the promise of two bottles of beer each 'when the overall situation resolved itself'. The young soldiers, who were impatient youths and perpetually hungry, fastened on their sandwiches avidly the second they appeared; the old soldiers stowed theirs carefully away in their packs in anticipation of a leisurely and beer-soaked repast at some later date.

The fusiliers of 'Z' Company, with or without turkey under their belts, were ready.

On the night of April 24th, the Gloucesters were gradually being hemmed into an ever-tightening perimeter; the Belgians were beating off one attack after another; the Royal Ulster Rifles, the reserve battalion, waited for a chance to avenge the January massacre. On this night the Chinese hurled themselves on 'Z' Company of the 5th Fusiliers, and they came from all sides. It was a night of furious and unremitting fighting and it went on for nine mortal hours.

The first weight of the attack fell on Second-Lieutenant Shepherd's platoon. Handling his platoon and controlling his fire like a veteran, Shepherd beat off successive waves of Chinese who surged to within a few feet of his position – advancing in short rushes, taking cover, dodging, hurling grenades, firing burp guns. When they finally drew off for a merciful respite, the Chinese called upon Shepherd's men to surrender – 'Surrender, Englishmen!' they shouted in fractured English; 'surrender, and don't be killed!' Shepherd's men replied to these exhortations with long bursts of fire, raucous laughter by numbers, and vociferous instructions.

John Winn knew that any more similar assaults must surely overwhelm Shepherd's position. On the ridge where 'Z' Company were dug in were two huge rocks, towering and menacing on the skyline, and these rocks were vital to their lives and their defence. Shepherd's men held the southernmost rock, while the other was temporarily unoccupied. Major John Winn saw this rock, and his dominant thought was that it must be held absolutely against all comers. There were few enough men available – casualties were mounting from the rain of bullets and grenades. Winn came to a decision: he would hold this vital rock himself, with the considerable assistance of Fusiliers Tinsley and Topliss. It was this decision, perhaps more than anything else, which was instrumental in the admission of John Winn to a Companionship of the Distinguished Service Order: there are many 5th Fusiliers today who contend that his DSO was very nearly a Victoria Cross.

Prior to the occupation of his rock – the rock which he was to hold for the remainder of that fearful night – Winn had appeared to be everywhere at once. Almost irritatingly casual under murderous fire, he strolled nonchalantly from one section to another, cheering and encouraging his fusiliers in what Second-Lieutenant Paddy Baxter, an irreverent subaltern, described as his 'hunt ball' voice. And every fusilier who heard that voice resolved to fight that little bit harder, shoot that little bit straighter and hold on for that little bit longer.

Now Winn, Tinsley, and Topliss had taken up positions on the rock: Topliss had a sten gun and Tinsley a rifle, with which he was a deadly marksman. Winn started the action with an American carbine, a weapon which he liked better than any other, except possibly the shot gun which reposed with the baggage at rear echelon. But a grenade burst practically at Winn's feet, splintering the carbine to matchwood and wounding him in the right arm – the same arm which had been shattered in Italy seven years earlier. This indefatigable trio had, however, a plentiful supply of hand grenades, which they threw with unerring force and accuracy at the Chinese as they strove to close in on the rock.

Winn, Topliss, and Tinsley took an extravagant toll of the Chinese swarming round the rock, and one Chinaman met a singularly macabre end: creeping to within a few yards of the position, he was neatly dispatched with a shot from Fusilier Tinsley's by now almost red-hot rifle; the bullet struck a phosphorus grenade attached to the Chinaman's belt and he was promptly transformed into a human torch – John Winn recalls that the corpse smouldered gently for more than an hour and, appropriately, gave off an indescribably horrible kitchen smell: there were no turkeys roasted that night, but that particular Chinaman was done to a turn . . .

What was really needed, Winn quickly realized, was a bren gunner dedicated to the grim task of killing at close range, for one or two more determined rushes by the Chinese would surely swamp them.

Suddenly Winn bethought himself of Fusilier Crooks – Crooks, the man who declined to dig, put up barbed wire or peel potatoes; Crooks, the fighting soldier with the Military Medal; Crooks, the company nuisance. If Crooks was as good a fighting soldier as he claimed to be, there was ample opportunity for him to show his prowess on this rock.

During a brief lull in the fighting – ten minutes represented a long one – Winn bawled down the hillside: 'Pass the word for Fusilier Crooks!'

Crooks was in a slit trench, two feet and six inches deep, which someone else had dug, peering into the blackness and searching for a worthwhile target for his bren gun, to the accompaniment of blasphemous threats against all things Chinese. So far he had not seen one.

Quoth Sergeant Jim Gascoigne, who was no friend of Crooks: 'Company Commander wants you – reckon he's going to give you that fourteen days he owes you. Move yourself.'

Staggering under the weight of his bren gun and as many magazines as he could carry, Crooks made his way up to the rock.

'Private Crooks, sir,' he reported to Major Winn.

'*Fusilier* Crooks,' corrected Winn, pausing in the work of applying a field dressing to his arm; 'you're the very man I want to see. There are a lot of Chinamen out there, and I'm fed up with them. Let's see what you can do about it.'

'Yes, *sir*,' said Crooks, a fusilier whether he liked it or not, and promptly became a one-man wave of destruction.

He fired long bursts at long range and short bursts at a range of a few feet; he fired kneeling and lying down; he stood, silhouetted on the skyline, firing from the hip. Just as dawn was breaking he stood up on the rock, in full view of the Chinese, looking for fresh targets. A bullet struck him in the left shoulder; Crooks fell and rolled thirty feet down the hill.

But the Chinese had not seen the last of Fusilier Crooks. He crawled back up the hill, and continued to engage the still-swarming enemy from a lying position. When a merciful dawn broke, he was still there, bleeding profusely, and still firing.

The major from the Manor House near York and the fusilier from a mean street in Leeds had held the rock.

Meanwhile, CSM Radcliffe, known to the fusiliers as 'Old Mahogany Head' and hitherto untried in battle, had been a tower of strength throughout the night. With calm precision and heedless of flying bullets and shrapnel, he maintained a constant supply of ammunition to the forward sections. When Sergeant Clarke was killed, Radcliffe took over command of the platoon and under his leadership it yielded not one foot of ground, be the Chinese ever so numerous and the night ever so long.

By dawn of the 25th the Chinese had clearly had enough: for the loss of Sergeant Clarke and two fusiliers killed and twenty-eight wounded, 'Z' Company had, at a conservative estimate, killed five hundred Chinese and almost certainly wounded twice as many again.

The ubiquitous Fusilier Fred Wappit, the company medical orderly, was applying still more field dressings to the company commander's seemingly indestructible body; John Winn had been hit twice more during the night, in the head and thigh. Fusilier Crooks lay on a stretcher, smoking a cigarette provided by his company commander – like Montgomery, Winn was a non-smoker himself, but always carried an inexhaustible supply for his men.

Crooks said to Major Winn: 'I take back what I said, sir. The Northumberland Fusiliers isn't such a bad old mob after all.'

'Nice of you to say so,' said Winn weakly. 'We're glad to have you with us. I wonder if you'd care to accept the rank of fusilier?'*

Many of the fusiliers of 'Z' Company, furiously engaged all that night, were no strangers to war – Major Winn, Sergeant

* For his gallantry Ronald Crooks received the immediate award of the Distinguished Conduct Medal.

Clarke, Fusiliers Crooks, Tinsley, and Topliss had all experienced the agony of waiting, the clammy hands, the straining of the mind to a screaming tension which is the inevitable prelude to battle.

But on April 24th, twenty boys of the East Yorkshire Regiment became men, and by the same token 5th Fusiliers, in one short day.

They were National Servicemen, all aged 19, stationed in the reinforcement camp at Kure in Japan, not one of the more sought-after stations, waiting for the order which would send them to Korea as reinforcements for the 1st Royal Northumberland Fusiliers; they did not like the idea, and said so with mournful truculence.

The order came all too quickly: a peremptory summons to the camp orderly room on the evening of the 22nd told Sergeant Douglas Gardiner that he was to take a draft of green, bewildered and largely unwilling youths to Korea forthwith. Sergeant Gardiner was not unduly perturbed – seven years in the British Army had long since taught him not to be surprised at anything, and he hastily assembled his audibly discontented flock. They were flown to Korea in an aeroplane relentlessly stripped of all comforts. They rode in jolting discomfort up a road, with Chinese swarming in the hills on both sides. A burst of machine-gun fire partially severed the head of one young National Serviceman and wounded two more. By the time they reached Battalion Headquarters and came under the morose strutiny of RSM Richardson (who had heard many things about National Servicemen, most of them discouraging) they were very nearly men, but not quite.

On 'Z' Company's position they saw and smelt the effects of total war at close quarters: fusiliers exhausted and grimy; fusiliers badly wounded; fusiliers dying; fusiliers dead. They saw and smelt the piles of Chinese corpses heaped round the positions. They observed, at a very early age, that war was filthy, sordid, and wasteful.

Their job was to help the wounded of 'Z' Company off the hill, and they set to with a will, just as Major Robin Leith-MacGregor's 'Y' Company arrived to effect a relief of the position. The Chinese attacks had ceased for the time being, but snipers were busy. And as these boys spoke the comfortable words of the soldier – 'lean on me, mate', 'have a draw on this fag' and 'take it slowly, mate' – to suffering men whom they were meeting for the first time, they heard the spiteful whine of bullets and thought of young Jim whose head had suddenly become a red mess, and so they became men . . .

Major John Winn was the last down; he had refused to be evacuated until the last fusilier, wounded or unwounded, was off the hill. He walked down unaided, except for a thick ashplant, as he had walked his estate in Yorkshire in search of partridge.

Winn was met at the bottom of the hill by Major Miles Speer, second-in-command of the battalion. To Winn he said in tones of mild reproof: 'Good morning, John. You're looking extremely untidy.'

Winn lowered himself gingerly on to a conveniently sited ammunition box, and accepted a mug of strong tea, generously laced with rum. Before he could take the first warming sip, he unobtrusively and, so it seemed, apologetically, lost consciousness . . .

[7]

In the Battle of the Imjin the gunners of Lieutenant-Colonel Maris Young's 45th Field Regiment and Major Terence Fisher-Hoch's 170th Independent Mortar Battery did a magnificent job.

Morning, noon, and night the sky was filled with the snarling whine of shells, tearing down the last veils between the Chinese and hell, and sweating gunners rammed shells into the open breeches as fast as they could pick them up. It seemed that nothing could live in the ceaseless stream of shells, but still the Chinese came on . . .

Neither gunners nor infantrymen could kill them fast enough, and it soon dawned on commanders at all levels that this was a battle in which boundless courage and tactical superiority meant less than nothing: it was a battle of *attrition* – for each man killed there were ten replacements; in spite of their losses, the strength of the Chinese was not merely constant but increasing all the time.

The 45th Field Regiment had twenty-five-pounders and they were firing more rounds per weapon – the average was put at a thousand – than had been fired at El Alamein, theretofore considered the biggest show ever put on by British Artillerymen.

They dropped shells on Chinese less than thirty yards from the forward companies' positions; they shelled enemy massing on the far river bank and enemy on the way across, and hundreds of Chinese were screaming and dying in the Imjin which was running red with blood. The guns traversed full circle: they had a range of thirteen miles, but during this battle they were oft times firing point-blank, over open sights, at Chinese soldiers a mere fifty yards away. Towards the end of the battle, every shell in Korea had been delivered to the gun positions, and the exhausted gunners were almost too tired to lift them. At two airstrips lorries were waiting for a fresh supply of shells which would be rushed up to the waiting guns.

The United States Air Force, too, performed prodigies of valour in the Imjin battle. On April 23rd, St George's Day, there were three air strikes in support of the Gloucesters, Belgians, and Northumberland Fusiliers. On the 24th, an American air force officer – a conscientious young lieutenant of a statistical turn of mind – attempted to keep an accurate count of sorties, but eventually gave up the unequal struggle: like the gunners, the American pilots answered every call for help – they machine-gunned, rocketed, and seared with napalm Chinese soldiers scattered round the flanks of the Gloucesters' and Northumberland Fusiliers' positions. In the panic caused by one of these napalm attacks seven Gloucesters, who had been taken prisoner the day before, first watched with awed fascination the roasting alive of their captors and then made a break for it down the hill to the Fusiliers' lines.

Pilots of helicopters engaged in the evacuation of casualties displayed almost suicidal heroism: a fusilier, shot through both lungs, was picked up fifty yards away from the Chinese and was on an operating table within half an hour of being hit; he lived to tell the tale in his favourite Newcastle pub on his return home.

But the Chinese, like a huge, amorphous mass, still came on . . .

[8]

As dawn broke on April 25th it was clear to Brigadier Tom Brodie that he had no choice but to withdraw his brigade: a counter-attack, with all his forces committed to desperate defensive action, was clearly out of the question; the Gloucesters were completely surrounded, and the penetration in the western sector was so deep that encirclement of the whole brigade was threatened.

For Tom Brodie it was a bitter decision, but he had been a Chindit in a war where bitter decisions were taken every day: he faced the problem of having his brigade very severely mauled, and of probably losing the Gloucesters altogether. The Belgians had been extricated with comparative ease; they had suffered severely, but not anything like as severely as the Chinese who had opposed them, and almost literally firing over their shoulders they had reached safety. The extrication of the Northumberland Fusiliers and Ulster Rifles seemed certain to be a long drawn out and bloody business. The withdrawal of the Gloucesters was an impossibility, unless some miracles intervened – and miracles in the Battle of the Imjin were at a cripplingly high premium.

Brodie's heart was heavy within him: the 5th Fusiliers and Royal Ulster Rifles must get out of it quickly, and would undoubtedly suffer severely in the process, but there was only one order that he could give to the Gloucesters and that was 'Hold on where you are'.

Brodie spoke personally to Colonel Fred Carne on the wireless; a man not given to displays of emotion, he nevertheless sounded anxious and strained and with good reason: he was committing the Gloucesters to a desperate task which only they could carry out; no relief force could be sent, but every effort would be made to dispatch one on the morrow. Meanwhile, they must hold and that's all there was to it.

Colonel Fred Carne, his pipe still burning evenly and unhurriedly, heard the brigadier out in sympathetic silence: between these two men existed a mutual liking and respect; they trusted one another implicitly. Brodie knew that Carne would hold out as long as there was breath in his body; Carne knew that Brodie would spare no effort in sending help if it was within his power. Meanwhile, the Gloucesters were just three hundred strong; the wounded lay exposed in the open, and many of them had been re-wounded by the lashing machine-gun fire that hardly let up for half an hour at a time; there was just enough ammunition for another twelve hours' fighting; there was no food and little water.

Weighing up all these defects, thumbing a new fill of navy cut from a fast-dwindling supply into the blackened bowl of his pipe, Carne spoke into the wireless: his voice was as steady and unemotional as if he had been on manoeuvres on Salisbury Plain.

Carne said: 'I understand the position quite clearly. But I must make it clear to you that my command is no longer an effective fighting force. If it is required that we shall stay here, in spite of this, we shall continue to hold.'

There was nothing more to say. Brigadier Brodie said to his Brigade Major, Ken Trevor, a Cheshire like himself, 'All right, then – it's time to go . . .'

The orders to withdraw were not well received by the Royal Ulster Rifles. They had levelled the score on the January ambush at an encouraging rate of interest: during the forty-eight hours of battle they had yielded not one inch of ground, and had severely punished every enemy attempt to penetrate their position; Major Sir Christopher Nixon, commanding 'A' Company, and a man not given to exaggeration, claimed to have beaten off no fewer than sixteen attempts to oust his men from their position. Exhausted and sadly depleted in numbers, as they were, the riflemen had every confidence that they could continue to hold the position and fight the Chinese attack to a standstill.

The orders for the withdrawal were that the Royal Ulster Rifles and Northumberland Fusiliers should break off the action and withdraw immediately down the road, through a block formed by 'B' Company of the Ulsters, and thence south-east to the main Uijongbu road; further defensive positions would then be occupied just north of Seoul.

The withdrawal had to be carried out speedily: the lateral valley in which Brigade Headquarters lay was being harassed by small groups of enemy, which would very soon become large ones; the guns of 45th Field Regiment were under small-arms fire and were engaging the enemy over open sights.

The extrication of the Fusiliers and the Rifles was achieved, but only with fearful effort; in pulling out, the two battalions suffered still more heavy casualties. They would have been considerably heavier if it had not been for the extraordinary gallantry and determination of Major Henry Huth's 'C' Squadron of the 8th Hussars.

So far in Korea the 8th Hussars had had little opportunity to shine. But once the Battle of the Imjin got under way in real earnest, the tanks, like the artillery, were in perpetual demand; invoked like guardian angels at every point of threat and crisis – which was almost everywhere. As soon as it became apparent that the infantry were in grave danger, the 8th Hussars went into action with the same *panache* and reckless contempt for danger that they had displayed at Balaclava.

There was no easy way back for the 5th Fusiliers and the Royal Ulster Rifles: they came down the road, in full view of the enemy on both sides of it – a road which was ceaselessly swept by fire. In the words of Henry Huth, it was 'one long bloody ambush'.

For the Ulsters it was the January massacre all over again, only more so: in January there had, at least, been a merciful darkness. But this ambush was in broad daylight, which meant that the Chinese could select their targets with care, and on this fearful march the targets were many and gratifyingly easy to hit.

Throughout this hellish maelstrom, the 8th Hussars did a truly magnificent job, and raked the surrounding hills with fire as the battered remnants of the Ulsters and Northumberland Fusiliers struggled down the road. Displaying either almost superhuman courage or extraordinary stupidity, many Chinamen launched frontal attacks on the tanks: due to the ferocity of the fire brought to bear on the road, hatches were closed down and the lingering fog caused periscopes to become misted to such an extent that the tanks were frequently driving blind.

One Chinaman debouched from a ditch and hurled a sticky bomb slap against the side of Lieutenant Michael Radford's tank; it exploded, blowing a gaping hole through the skirting plate, but leaving the suspension undamaged. Radford took swift and ruthless counter action and crushed the intrepid bomb-thrower under his tracks. Sergeant Jack Cadman heard an insistent battering on the turret over his head; he directed the tank straight into a house, brushed the Chinaman off and ran over an enemy machine-gun post beside the road. Captain Peter Ormrod and Captain Gavin Murray both had Chinese clambering on to their tanks; showing a heartening disrespect for War Department property, they shot at one another, killing a dozen each.

For the riflemen and fusiliers the choice lay between walking and riding on the tanks; the walkers fared the best, for at least they could take advantage of the scant cover provided by shallow ditches at the side of the road. But many others – older and lazier men, who considered that they had done enough marching – climbed on to the tanks and were shot off them like flies. Running that gauntlet of fire, the wounded experienced the worst ordeal of all: many were re-wounded as they lay on and around the tanks' turrets; some, too weak to hold on, fell into the bullet-swept road and did not get up again. Horribly exposed, the wounded could do nothing to ease their frightful travail as the bullets thudded into them, and the tank crews were almost as impotent: the tanks were so crammed with bodies and slippery with blood that it was impossible to traverse the guns. Among the many dead was the Commanding Officer of the Northumberland Fusiliers, Colonel Kingsley Foster: he had been the last to leave the battalion position, and his jeep received a direct hit from a Chinese mortar bomb.

In this manner the Royal Ulster Rifles and Royal Northumberland Fusiliers came out of battle.

TWELVE

FOR the Gloucesters there was to be no withdrawal.

As the bone weary, wounded, dead, and dying 5th Fusiliers and Ulster riflemen came to Uijongbu, Colonel Carne and the three hundred Gloucesters still on their feet made ready for their last fight.

Clearly, they had all been there far too long: this was at once apparent from the staring red-rimmed eyes; the gaunt and wasted faces covered in thick and filthy stubble – there was hardly any water for drinking, let alone washing and shaving; from the trembling hands which now had but one function – to press a fresh clip of ammunition into a rifle's magazine or a new magazine on to bren or sten gun. But there was little enough ammunition left, the wireless batteries were practically exhausted and their last square meal had consisted of one hard boiled egg and a biscuit of surpassing hardness just twelve hours earlier.

Yet it seemed that relief was imminent. The night before Captain Anthony Farrar-Hockley had spoken on the wireless to the Brigade Major, Ken Trevor: a relief force in the shape of Filipino infantry and tanks of the 8th Hussars was on the way and should be with them by dawn on the 15th. Farrar-Hockley recalls that the Brigade Major sounded cheerful and optimistic: 'We'll have a drink together tomorrow evening,' he told Trevor, 'and you'll be paying.'

But now it was ten o'clock, the Chinese were massing for further attacks and there was no sign of any relief force.

There never was. Three Filipino light tanks, which were leading the column, were knocked out by mines and effectively blocked the road to the vehicles behind. The Centurions of the 8th Hussars, operating in country where no Centurion had ever operated before, could advance no farther. The infantry, pitilessly exposed and pinned down by fire, could make no progress. This was the first relief column, and neither of the other two, composed of Belgians, Puerto Ricans, and Americans, made any better headway.

The Gloucesters were now finally and irrevocably on their own.

Lance-Corporal Bob Matthews was in the throes of a complicated calculation of ballistics, but he was too tired to get it right – every time he did the multiplication sum he got a different answer.

He had a full magazine on his bren gun and two more to hand. There were twenty-eight rounds to a magazine, and assuming that he killed a Chink with each round, how many dead Chinks would that make? He had done the sum six times already, and each time had arrived at a different answer – sixty-eight, eighty, ninety-two . . . suddenly it seemed desperately important to get it right.

'Harry!' he called to the soldier on his right, Private Harry Johnson.

'Ain't got none,' replied Johnson hurriedly, anticipating a further dwindling of his scant supply of cigarettes.

'What you mean, you ain't got none? I want to know what three times twenty-eight is.'

'Christ knows,' said Johnson, to whom education certificates represented daunting and as yet uncrossed hurdles. Johnson had a problem of his own; he had exactly five rounds left in his rifle . . .

Understatement . . . stiff upper lips . . . the British cinema industry had always made big play with these essentially British ingredients. It goes without saying that the 1st Gloucesters, even in the desperate circumstances in which they found themselves, produced their own crop.

Racing enthusiasts have heard the casual and well modulated voice of Tony Preston on BBC television, extolling the virtues and defects of horses as they sedately circle the paddock: in the Battle of the Imjin, Second-Lieutenant Preston's platoon was in one of the more exposed parts of the perimeter and exposed to ceaseless attack. During one furious encounter he interrupted a telephone call to Farrar-Hockley saying: 'Excuse me, old boy, I must hang up now – they're only ten yards away.'

Anthony Farrar-Hockley, not a soft-spoken officer and one accustomed to being obeyed, issued this order to Lieutenant 'Guido' Temple, of Gloster Crossing ambush fame: 'Guy, you will stay where you are until further notice. If your ammunition runs out, hurl bloody rocks at them.'

Colonel Carne, encountered by his Adjutant with a rifle and fixed bayonet slung over his shoulder, explained this unusual circumstance by saying: 'Just been shooing away some Chinese.' (He had, in fact, collected a scratch force consisting of CSMI 'Muscles' Strong, two regimental policemen, and a company clerk and bayoneted and grenaded a marauding party of Chinese off the position.) A frantic telephone call from an American general wrung from Carne the admission, 'Well, it *is* a bit sticky.'

[2]

There was no help devised by man for a unit so desperately stricken as the Gloucesters: the term 'doomed battalion' savours of the melodramatic, and is a phrase normally only dreamed up by Hollywood scriptwriters. Nevertheless, the Gloucesters were doomed, and every officer and man knew it. But they were not prepared to admit it – not quite yet . . .

They were desperately short of food, water, and ammunition, and their sense of isolation was fast becoming acute: they asked for helicopters to evacuate the wounded, but no helicopter could land amid the curtain of steel which enveloped Gloster Hill; supply drops were attempted, but the exhausted men in the trenches could only watch in despairing impotence as precious supplies fell inaccessibly on to the lower slopes, where still more Chinese massed for the final annihilating assault. Even the wonderful support they had received from 45th Field Regiment was showing signs of slowing down: the gunners had more than enough troubles of their own, and were firing over open sights at Chinese intent on swamping the gun positions.

During the night of April 24th–25th, the rifle companies, all reduced to barely a platoon strength, beat off successive enemy attacks: the Chinese had declined to learn from previous mistakes, and with their vast reserves of expendable manpower they had no need to – they continued to come at the Gloucesters head-on, and as if to signal imminent victory their bugles maintained a hideous cacophony of discordant sound.

'B', 'C', and 'D' Companies were all under ceaseless attack during that last night of furious fighting, but Majors Denis Harding, Paul Mitchell, and Mike Harvey continued to send in reports to Battalion Headquarters – the kind of reports that Fred Carne, knowing his men, expected to receive: 'They're all round us . . . afraid we may have to pull off this hill . . . getting very short of ammunition . . . doubt if we can hold 'em off for more than an hour . . . you're very faint – battery must be running down . . . not a hope of mounting a counter-attack . . . we're down to twenty men, and ten of them are wounded . . .' They were the clipped and economical voices of men educated at the less expensive public schools and the Royal Military College, Sandhurst: the voices of men who had learned duty and discipline the hard way; men who had been brought up to face reality, and they were facing it now. But one thing was certain: they would not be able to keep it up for much longer.

Among Gloucesters who fought in this battle, it was universally agreed that 'A' Company had the roughest handling of all: the company commander, Major Pat Angier, had been killed, as had VC winner Lieutenant Phil Curtis, and Second-Lieutenant John Maycock; Lieutenant Terry Waters, severely wounded in the head, was only prevented from going back to his platoon by the physical intervention of Captain Bob Hickey, the medical officer.

The remnants of 'A' Company – some thirty men in all – had been forced off their position: they had not abandoned it, because Gloucesters have never abandoned any position since their inception in 1694, and have forty-five battle honours to prove it; they had literally been *blasted* off it, for the shallow trenches which they had been occupying had afforded no cover whatsoever, and they might just as well have been lying out in the open.

To this still belligerent but leaderless force came the Adjutant of the Battalion, Captain Anthony Farrar-Hockley; the men in the trenches experienced a brief spasm of apprehension at his approach, for the Adjutant was something of a fanatic on the subject of discipline and soldierly bearing. A brief ripple of conversation went up and down the men in their shallow trenches:

'Eyes down, 'ere comes the Fuehrer . . .'
'Better put your hat on straight . . .'
'Cigarettes out everywhere . . .'
'Bet he checks me for a f— haircut . . .'

Farrar-Hockley came to 'A' Company, because Battalion Headquarters, as a strictly military entity, had virtually ceased to exist: the Commanding Officer was almost certainly strolling unconcernedly among the men in the forward positions, if he was not actually leading a section with rifle and bayonet (as it happened he was); RSM Jack Hobbs was undoubtedly somewhere where the fighting was thickest – replenishing water and ammunition, deriding the faint-hearted, joking with the company sergeant-majors, assisting the wounded, dishing out fags, checking men for haircuts, threatening men with stoppage of forty-eight-hour passes; the signals officer, Captain Richard Reeve-Tucker was dead; the Assistant Adjutant, Lieutenant Donald Allman, had long since abandoned all administrative chores; ignoring a painful wound in the shoulder, he was commanding a platoon composed of a dozen men. Tony Farrar-Hockley, winner of the Military Cross in Italy and a fighting infantryman to the bone, was not going to sit on the end of a wireless set with a fast-failing battery, waiting for messages which never came when there was urgent work as a fighting man waiting for him.

He could not have come at a better time: he strolled nonchalantly up and down the line greeting those men that he knew by name: he saw Privates Bill Masters, Sid Middleton, and Bert Guilding, three reservists who, in Colchester, had produced reasons of a startling complexity but amazing similarity for being sent home immediately; he saw two of the thirstier members of the Sergeants' Mess, Jim Clayden and Tom Tuggey; he saw young Private Ted Fish, his head bandaged, scrounging a fag off his neighbour without conscience or shame. He saw, at a glance, that these were good men to be in a tight corner with.

Here and there he paused to crack a joke: they were not particularly funny jokes, but Tony Farrar-Hockley knew that they were the sort of jokes the men must have: some were lavatorial and others just plain crude, but they raised a laugh or two and that was all that mattered.

Then, suddenly, Farrar-Hockley was all the soldier, as he pointed to the hill which 'A' Company had just vacated. 'Now, then,' he said crisply, 'who's going to win the VC? We're going back there, and that means all of you.'

With the Adjutant at their head, the men of 'A' Company charged across thirty yards of open and bullet-swept ground. Now they were in among the Chinese: triggers were pulled at a range of a few inches; rifle butts cracked down on Chinese heads; bayonets carried out the manoeuvre 'in – out – on guard' with barrack-square precision. 'A' Company was back on its original position.

[3]

Captain Bob Hickey, the medical officer, worked ceaselessly and probably had less sleep than any other man on Gloster Hill. On April 25th, the stretcher cases, numbering some eighty men in all, lay in the open: men with lower jaws shot away, fearful cavernous wounds in stomach and chest, fingers, arms and legs missing, faces a red soggy mess. There were men re-wounded as they lay in the open, and men like Sergeant George Robinson and Private Albert Robson, the over-enthusiastic bren gunner of 'B' Company, who refused to be evacuated and continued to fight until they lost consciousness.

Without sleep and with a miserable dole of drugs, Hickey tended the wounded with a dedication which, in the annals of the Royal Army Medical Corps, may have been equalled but surely had not been surpassed: by day in full view of the enemy and under a hail of bullets; by night amid an inferno of hellish noise, with only the dying battery of an electric torch to guide him.

Mercifully one of the more successful airdrops delivered

plasma and blood transfusion, without which many of the wounded would have died; without Bob Hickey many more of them would have.

Usually in close proximity to the medical officer and labouring with the same selfless devotion, was Sam Davies, the Chaplain. He had seemingly inexhaustible supplies of cigarettes: like Private Charlie Edkins, he had not wasted his time in Pyongyang during the big 'bug-out'. Throughout the battle, Davies walked an incalculable number of miles, cheered the living, succoured the wounded, comforted the dying and buried the dead.

Sergeant Jack Eames, Second World War Military Medallist, indestructible soldier of twelve years' service, had quaffed his last pint in the Sergeants' Mess, roared his last order in a parade-ground voice which was clearly audible at three hundred paces: for Jack Eames was dying – Sam Davies knew it, and Jack Eames knew it too. Bob Hickey paused to look at Eames on one of his errands of mercy, and then just perceptibly shook his head at Davies – there was a lance-corporal ten yards away, who might last the night if attended to promptly; Sergeant Eames, Hickey decided rapidly, had another hour at the most.

Eames talked to the Chaplain about his mother and his young wife; he had not been long married and was, as he shyly confided to Davies, just getting to like it. They prayed together for a while, and Davies could see that the end was very near; Eames must have known it too, for he said: 'I'll be all right, padre. You can leave me now.'

On his way back to the aid post, Davies came across a dead private, just 19 years old – 'too young to vote, old enough to die', thought the Chaplain in a mood of unwonted secular bitterness, and removed his rosary beads: these beads remained with him throughout two and a half years of captivity, and on his return he gave them to the young soldier's parents.

Private Jim Walker had always been an individualist to a marked degree; he was averse to taking orders, and frequently to carrying them out, a characteristic which tried the patience of his NCOs most sorely. He was to practise this spirit of independence to the last: ordered by Lieutenant Henry Cabral to take up position with his bren gun and engage a party of Chinese climbing the hill towards them, he ran straight down the hill at the enemy, firing the gun from the hip as he went.

To Colonel Carne, the noise of the Chinese bugles was nothing more than a minor irritant; to Tony Farrar-Hockley, a less tolerant and less patient man, it was a filthy bloody racket, perpetrated by a shower of slit-eyed, oriental unwashed yahoos who weren't fit to carry a Gloster latrine bucket. He stuck his head out of the slit trench which he currently tenanted and bellowed at the top of his voice: 'DRUM-MAJOR!'

'SIR!' shouted a voice, if anything louder, and the aggressively moustached features of Drum-Major Philip Buss appeared out of another hole in the ground.

'Got a bugle?' demanded the Adjutant.

'SIR!' blared 'Drummie', who was a man who never used two words when one was sufficient.

'Then play it, Drum-Major, play it!'

The Drum-Major knew the fitness of things: clearly, he could not play the bugle sitting in a slit trench. He stood, silhouetted in the half light, and proceeded to play every call he knew with exemplary perfection – Roland's horn among the mountains, gathering echoes from centuries of battle, sounded again.

He played the long Reveille ('Get out of bed, you lazy b—') and the short Reveille, universally known in the British Army as 'Charlie':

> Charlie, Charlie, get up and wash yourself!
> Charlie, Charlie, get up and pee!

He played 'Cookhouse' – 'Come to the cookhouse door, boys' – which raised an ironic cheer from the haggard men in the trenches, and 'Defaulters', which caused a number of men in 'Spike' Pike's platoon to wince visibly. He followed this with 'Orderly NCO calls', 'Officers Dress for Dinner', 'The Last Post', and 'Lights out'. Finally, as if his versatility had to be proved beyond all doubt, Buss produced the American version of 'Revelly' and 'Taps'. The Gloucesters cheered him to the echo.

It was a crazy occasion; the sweet notes of the bugle, the crash of mortar, rifle and machine-gun fire, the wild cheering of men almost certainly doomed to death or captivity.

Colonel Fred Carne carried them all.

Prior to going to Korea, Carne had made no startling impact on his battalion or, indeed, on anyone else: a somewhat taciturn, withdrawn and reserved man, in Colchester he had seemed to be a remote and slightly Olympian figure. His pre-war career, since joining the Gloucestershire Regiment from Sandhurst in 1925, had been unexciting, and his Second World War service unspectacular – he had spent much of the war on the staff and in East Africa.

But during the desperate Imjin battle he was, by common accord, magnificent, stupendous, superb, superhumanly brave, fearless – military chroniclers soon ran out of adjectives; it seemed that from inexhaustible reserves of strength and courage he gave every officer and man a share.

Throughout the entire engagement, showing a complete disregard for his own safety, the Colonel moved among the whole battalion, contemptuous of murderous and accurate fire, inspiring the utmost confidence and will to resist among his men. He knew that his flanks had been turned, and towards the end he knew that no relief force could possibly reach him. But he also knew that the abandonment of his position would clear the way for the enemy to make a major breakthrough and thus endanger the whole Corps; he inspired his officers and men, even the most recalcitrant and loudly 'ticking' reservist, to fight far beyond the normal limits of human endurance, in spite of overwhelming odds and ever-increasing casualties, shortage of ammunition, food and water.

He displayed, in the austere words of the official citation for his award of the Victoria Cross in 1953, 'powers of leadership which can seldom have been surpassed in the history of our Army'.

[4]

For the Gloucesters – they were later to be known as 'The Glorious Glosters', and the title is appropriate enough – the sands had all but run out. Ammunition was distributed for the last time to men still fit to bear arms: each rifleman had three rounds; each bren gun one and a half magazines, or forty-two rounds (Lance-Corporal Bob Matthews now knew exactly where he stood); each sten gun a half magazine. Some indication of the enormous quantities of grenades thrown at the enemy was revealed by a last count: there were just seven left in the whole battalion. Captain 'Spike' Pike had three rounds left for his pistol; Major Denis Harding, who had led his fifteen men of 'B' Company back to Gloster Hill the day before with four rounds in his revolver now had none, and no revolver either. He had, however, a walking stick and five cigarettes.

At breakfast time – breakfast was tinned fruit or condensed milk for those who were lucky, nothing for those who were not – the last message came from Brigadier Tom Brodie on the wireless, giving Colonel Carne discretion to break off the action and for every man to make his own way out as best he could. The brigadier, normally a man sparing with his praise, ended the message with the statement 'Only the Gloucesters could have done it', and it came from Tom Brodie's heart.

Carne called his last officers' conference. His voice was firm and steady, but his face was grave and the famous pipe was at last empty and cold (one of his last recorded sentences before the break-out started was, 'Any of you chaps happen to have a spare twist of tobacco?'). And rumour has it that he got at least one more ounce of assorted brands before the day was out.

It was, said the Colonel, a case of every man for himself; there would be no relief column, no more air strikes, no more artillery support, for the gun lines of 45th Field Regiment were under heavy attack themselves. 'Good luck,' said Fred Carne, 'and God bless you all.'

And it was only then that his voice faltered a little as he turned to Captain Bob Hickey and said: 'I'm afraid we shall have to leave the wounded behind.'

'Very well, sir,' said Hickey. 'I quite understand the position.' On this vexed question of administration the CO had no need to give an order, and he knew it, and Bob Hickey knew it too.

Chaplain Davies also elected to stay with the wounded – it had never occurred to him, or to anyone else, that he would do otherwise.

The more seriously wounded had no choice in the matter: accompanied by Bob Hickey, Sam Davies, and Sergeant Knocker Brisland, the medical sergeant, they lay on their stretchers, waiting for the Chinese. Every company of the battalion had had a 'whip-round' of cigarettes, and they were well stocked – even Charlie Edkins had parted with twenty.

But there were many other soldiers, all of whom had fought more than well, who were clearly of a somewhat dubious quality when faced with a desperate escape endeavour: some were slightly wounded, some unashamedly asleep (and who had better right to be?), some so utterly exhausted that they could hardly put one foot after the other or raise a match to light a cigarette. For several Gloucesters the aftermath of battle consisted of only the desire to eat and sleep again, and if it had to be in Chinese hands, then that was too bad.

Sergeant Joe Martin of 'C' Company had recently shaved; his gaiters fitted round his ankles neatly and tightly; his rifle, with 'one up the spout' and five more in his pouches, was slung over his shoulder. The light of battle was in his eyes, and he looked ready to go through the whole battle again if necessary. He had won the DCM at El Alamein, and one thing was certain – if anyone was going into a prison camp, it wasn't going to be Joe Martin. It was time, he considered, for one or two choice items from his vocabulary, and from an extensive repertoire he selected: 'You blokes are like a lot of old whores at a virgins' meeting. Now, then, let's be having you.'

Several men rose with alacrity, and none more than Private Fred Farraday: Farraday had been 'put in the bag' at Singapore – he had worked on the Death Railway, stood to attention for twenty-four hours in the pouring rain, had his bottom teeth kicked out in one of a Japanese guard's more playful moments and arrived back in England three and a half years later weighing a fraction over six stone. He wasn't having any more of that lark . . .

[5]

The variegated adventures of the Gloucesters in their escape attempts would fill several volumes: the remnants of 'A', 'B', and 'C' Support Companies went south, and immediately found themselves under heavy machine-gun fire: there is little doubt that the Chinese could have mown them down like grass before a scythe if they had wished, but displaying humanitarian tactics which were certainly not in evidence later, they did not do so.

The idea of surrender was repugnant to Captain Anthony Farrar-Hockley, but there was clearly no other choice if the men with him were to stay alive for another five minutes. Feeling that he was betraying everything that he loved and believed in, he called out: 'Stop!' and then 'Put down your arms.'

'Spike' Pike and his party found themselves similarly placed: Pike actually loosed off two of the four rounds remaining in his pistol at the Chinese milling all round him, but realizing that his men probably had less than fifty rounds between them and the Chinese numbered something like five hundred, he gave the same order as Farrar-Hockley. In Tony Farrar-Hockley and 'Spike' Pike the Chinese had two singularly fractious and unco-operative prisoners: Farrar-Hockley made the first of three escape attempts that very night.

Lieutenant Henry Cabral, always an individualist to a marked degree, set off alone: finding himself completely surrounded by Chinese, he decided to hide in a bush until darkness fell. But darkness did not come quickly enough, and he too was captured; Henry Cabral was to die in a prison camp, after faithfully 'adopting an incorrect attitude', as the Chinese neatly put it, and being a constant thorn in their sides for twelve months.

Colonel Carne has been erroneously reported in some publications as having elected to stay with the wounded. This would have been entirely foreign to his nature, for as a soldier he had one duty only: to evade capture if possible and to return to fight at the earliest possible moment.

Together with RSM Hobbs and CSMI 'Muscles' Strong, he evaded capture for twenty-four hours, as did Private Bob Graham (one of 'Spike' Pike's pioneers) and Private Ted Fox, both of whom readily volunteered to join Farrar-Hockley in his second attempt at escape.

Having seen the fate of the parties which took the southern route, Captain Mike Harvey – in command of 'D' Company and a few machine gunners, ninety-two officers and men in all – decided on another plan: he headed north, towards the Chinese rear, then swung west and finally continued south. Before setting out, he told the men, in not very subtle terms, that there would be no stopping on this march and that wounded men, footsore men, bootless men, tired men, lazy men and dead men would have to be left behind.

At first it seemed that the scheme was succeeding admirably, and for the first three miles they saw no Chinese. But their luck soon ran out: they came under heavy machine-gun fire from all sides, and in the space of five minutes Harvey's command was reduced by half.

Marching by night, hiding by day; too tired even to swear; stumbling along in extremes of exhaustion; diving into water-logged ditches and ditches filled with jagged stones when still more machine guns opened up on them – in this manner they came suddenly into open country and saw the American tanks.

Sobbing with relief and exhaustion, they staggered towards the tanks. But their troubles were not yet over; the Americans, mistaking them for Chinese, opened fire and six more Gloucesters fell. Mercifully the tank men – they had no idea that any friendly troops were in the area – realized their mistake and ceased fire after the first few bursts.

That evening the roll of His Majesty's 28th Regiment of Foot, the 1st Battalion the Gloucestershire Regiment, was called. Of the 622 strong battalion which first went into action on the night of April 22nd, 1951, five officers and forty-one rank, and file remained.

[6]

This is the end of the story of the Battle of the Imjin, for which the 1st Gloucesters and 'C' Troop of the 170th Independent Mortar Battery were both awarded a United States Presidential Citation: today, any man who served in either of these units is entitled to wear the Citation's emblem – a strip of blue watered silk ribbon inside a frame of gold silk, on both upper sleeves of their uniform. It has been called the greatest and bloodiest fight of the entire Korean War, and few soldiers of any nationality will quarrel with this description.

The price paid by Brigadier Tom Brodie's 29 Brigade was a grievous one: the Gloucesters were virtually wiped out, and the total figure of killed, wounded and missing in his three battalions was in excess of one thousand. Hundreds more died, in varying circumstances of tragedy, in prisoner-of-war camps.

The Chinese casualties were enormous, the best and most practical tribute to the soldiers of 29 Brigade. Although they have never been published in strictly reliable form, the estimate of 10,000 killed and wounded, out of the 60,000 Chinese who attacked the Imjin line, can be taken as a fair one and, if anything, an understatement.

The Battle of the Imjin was more than just another battle honour for the Gloucesters, the Royal Northumberland Fusiliers, the Royal Ulster Rifles, and the Royal Regiment of Artillery; it did considerably more than merely provide one of the finest natural killing grounds for the British Army in any war.

The stand of 29 Brigade contributed more than anything else to the complete failure of the Chinese spring offensive, for not long after Colonel Fred Carne and his men disappeared into insubordinate and consistently uncooperative captivity, the Communist push on Seoul was called off. It was certainly the most influential single factor in the dashing of Communist hopes of celebrating May Day in the Korean capital. It proved beyond all doubt that mass assaults by expendable Chinese, however fanatically brave they may be, must always be doomed to failure against men like Tom Brodie's. Fred Carne, Anthony Farrar-Hockley, John Winn, Reggie Pratt, Edward Radcliffe, Ronald

Crooks, Jack Hobbs, Mike Harvey, Denis Harding – these are just a few of the men who are living proof of this.

There was a little city, and few men within it; and there came a great king against it, and besieged it, and built great

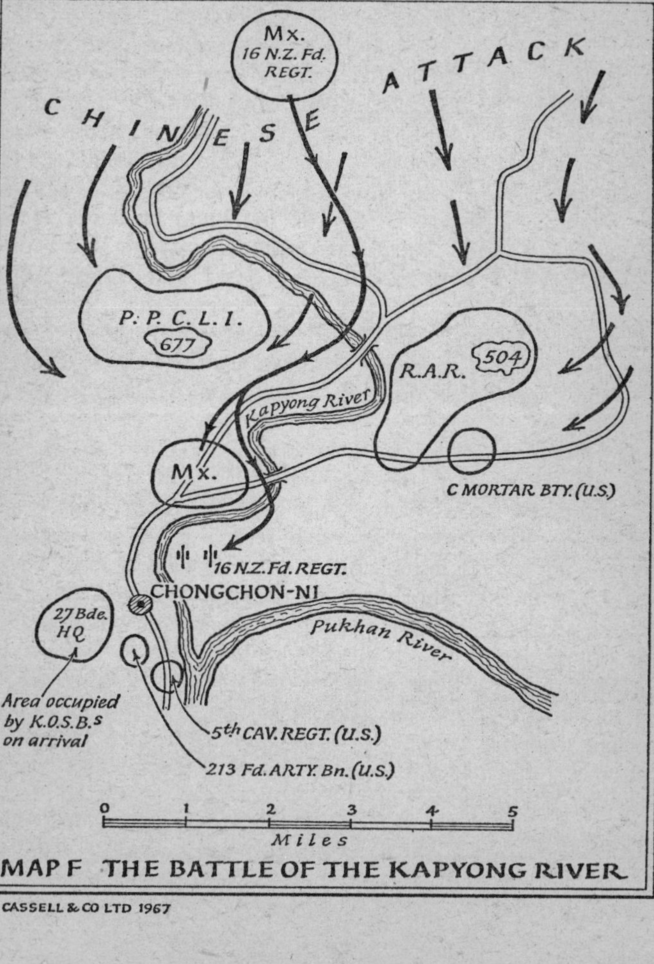

MAP F THE BATTLE OF THE KAPYONG RIVER

© CASSELL & CO LTD 1967

bulwarks against it:

Now there was found in it a poor wise man, and he by his wisdom delivered the city; yet no man remembered that same poor man.

These beautiful words from Chapter 9 of the Book of Ecclesiastes were taken by the 1st Gloucesters' new Chaplain as the text for his sermon at a memorial service. Captain Mike Harvey and the forty-five other survivors were there. Lieutenant-Colonel Digby Grist – he had succeeded Colonel Fred Carne in command, and had himself been shot in the wrist in the battle – had sent a signal to Brigade Headquarters, and its text was brief and to the point: 'We are operational again.' Looking at the young, pink and freshly shaved faces of the five-hundred-odd replacements, Digby Grist remembered other faces, young and many of them not so young: Fred Carne, Tony Farrar-Hockley, Pat Angier, Philip Curtis, Terry Waters, Henry Cabral; Sergeant Jack Eames, Drum-Major Buss, Peglar, the Provost Sergeant; Bob Hickey, the doctor, and Sam Davies, the parson; 'Spike' Pike and young John Maycock; Jack Hobbs, Walker, Robson, Fish, Edkins . . .

'In spite of this,' continued the Chaplain's address, 'the poor wise man who saved the city was very soon forgotten. In England they'll remember the Imjin battle for a little while – the soldier does have his day. I want to remind you this afternoon that it is not enough to remember now. We've got to show what we think of their sacrifice in the way we conduct ourselves in the days ahead . . . we must hand on to future generations some of the heritage of the past. Having handed it on, we will be in some measure worthy of those who died that we might live.'

The bugles played the 'Last Post', and then the 'Long Reveille'; the survivors of the battle, who had heard 'Drummie' Buss play on Gloster Hill, reckoned that these blokes were nearly as good – but not quite.

The new Gloucesters sang the National Anthem. Then the parade dispersed; men picked up their weapons and rejoined their companies.

The Gloucesters were operational again . . .

THIRTEEN

At the same time as the Gloucesters were still bowling over Chinamen like ninepins and the Royal Northumberland Fusiliers and Royal Ulster Rifles were carrying out their fighting withdrawal, the 3rd Royal Australian Regiment and the 2nd Battalion Princess Patricia's Light Infantry were administering a salutary thrashing to the Chinese from which they were a long time recovering.

This was the Battle of Kapyong, and for their performance in this action the 3rd RAR and the 'Princess Pat's' received the same award as that awarded to the Gloucesters: the United States Presidential Citation. It was scarcely less hardly won.

The Australians had earned a rest, if anyone had: without detracting in the least from the performance of the Argylls and the Middlesex, Brigadier Aubrey Coad often said that the 3rd Royal Australian Regiment had always experienced the lion's share of fighting in the early days, and neither the Argylls nor the Middlesex have ever questioned this: 'If I made the Aussies advance guard,' said Coad, 'they ran slap into trouble; if I gave them the job of rearguard, then they copped it from the rear.' No one was complaining about this, least of all the Australians themselves, and they had won the admiration of both British battalions, and of everyone else of any nationality who had seen them in action: by this time Aubrey Coad had grown accustomed to being addressed as 'cocky' by a grimy, slouch-hatted Australian private with a cigarette end hanging from a pendulous lower lip. At least, thought Coad ruefully and gratefully, some of the officers occasionally addressed him as 'skipper' or 'boss'. In a war like this the numbers of Chinese killed were more important than the phylacteries and protocol of the parade ground, and the Australians' score to date was impressive enough, as were their casualties. Aubrey Coad, who had at first regarded the presence of an Australian battalion in his brigade with a degree of apprehension, would not have exchanged them for the Coldstream Guards, *Les Paras* or the United States Marines.

As the Battle of the Imjin flared, roared and finally died away, the Australians, in blissful ignorance of the Chinese spring offensive, or indeed of anything else save the bottles of beer which lay comfortably to hand in every tent, made the most of their rest period. But the occasion was tinged with sadness: 27 Brigade, of which they were proud to be an integral part, seemed to be breaking up – the 1st Argylls had left Inchon for Hong Kong, and were being relieved by the 1st King's Own Scottish Borderers; the Middlesex, also, were expecting early relief by the 1st King's Shropshire Light Infantry. The 'Aussies', however, felt only fleeting envy: their rest camp near the village of Charidae, north-west of Kapyong, nestled peacefully in a grove of chestnuts; there were still patches of snow in the hills, but the wild pink azaleas were in bloom and the trees were coming out in fresh green leaf. Generally, the Aussies rated it a 'dinkum' place and Private Eddie Eyre, who enjoyed drinking and sleeping more than most, declared that he would be happy to stay there for the duration of the war.

Anzac Day – April 25th – was approaching, and it was rumoured that a celebration of no ordinary proportions had been planned: this particular Anzac Day was to be an exceptional one, because for the first time in history the traditional enemies, the Turks, were camped close by and they were soldiers who were imbued with the party spirit to a marked degree: Brigadier Tom Brodie is able to confirm this; he attended a luncheon party at Turkish Brigade Headquarters which, he recalls, lasted for five hours.

But Anzac Day, 1951, was, like St George's Day for the Royal Northumberland Fusiliers, indefinitely postponed . . .

[2]

The South Korean Army had been virtually bled white: since the beginning of the war in June 1950 it had suffered more than 150,000 casualties. Now, green recruits, who had been issued with rifles before learning how to fire them, filled the ranks of its 10,000-man-strong divisions. But the heaviest losses, as in any army in any war, had been among the junior officers and NCOs; as a result, countless squads and platoons floundered about the front – brave, but utterly bewildered and leaderless. Obviously they were in no fit state to withstand the shock of an all-out Chinese offensive, and some of their less charitable allies questioned whether they wanted to.

But in fairness to the South Korean soldiers, it must be pointed out that their divisions were not generously armed for total war: they depended on the Americans for artillery support when they needed it (and the decision in this matter was invariably American); tanks were only assigned to them for special missions. When the South Korean infantrymen fought Chinese or North Koreans one-for-one, they performed creditably and more than held their own, but not when they were outgunned and outnumbered. When the Chinese spring offensive hit them they had neither artillery nor armour, and they were both outgunned and outnumbered. The inevitable happened: the 6th ROK Division wavered, fell back and finally broke, exposing the United States 24th Infantry and 1st Marine divisions to either side of it.

In the valley of the Kapyong River the 3rd Royal Australian Regiment, the 'Patricias' and the 72nd United States Tank Battalion prepared to stand their ground, as the South Koreans streamed ever rearwards and the Chinese pushed ever forwards. The 1st Middlesex, in a mood of gay relaxation, already making plans for Homeric nights out in Hong Kong and expecting the 1st KSLI to relieve them at any moment, were rushed up to reinforce the Australians and Canadians.

The 3rd RAR and elements of the 72nd United States Tank Battalion were assigned a position on a ridge covering crossroads and fords over the Kapyong River and one of its smaller tributaries. Some three miles to the left the 'Patricias' held similar high ground covering a ford across the Kapyong River. Between the Australians and the Canadians, accepting their sudden and unexpectedly new operational role with cockney stoicism, were the 1st Middlesex. They were, however, bereft of their old friends, the New Zealand gunners who were far forward and supporting the fast-retreating South Koreans.

As in the Imjin Battle, men were woefully thin on the ground: both the Australian and the Canadian battalions were holding two battalion fronts, and there were alarming gaps between platoons. There was nothing to stop the Chinese, if they came in sufficient numbers, from pouring through these gaps and cutting off both battalions completely.

The hours after nightfall on April 23rd were unnerving for the Australians, chiefly because of the incessant clamour of noise all along their front.

Captain Reg Saunders, the Aboriginal officer of the 3rd Royal Australian Regiment, confessed to a feeling of depression as he listened to the approaching clamour of the retreating South Koreans: he had heard a similar sound in Crete and Greece and earlier in Korea, during the 'bug-out' period; it was the undisciplined noise of defeated soldiers in disorderly retreat. The noises are depressingly similar in any campaign, but South Koreans in such circumstances were exceptionally vociferous – there were shouted questions which inevitably went unanswered, and shouted orders which were either disobeyed or went unheard; bitter swearing, and one did not have to be an oriental linguist to glean that these men were fed up with the war and wanted no further part in it; a mechanized pandemonium of blaring horns, grinding gears and uselessly whirring self-starters; the sound of steel helmets, rifles and equipment being thrown away with a sulky finality; prayerful wailings which sounded as if they came from souls in torment, as indeed they did; shots fired despairingly into the night, and the proximity of some of the bullets caused Australians to start swearing on their own account.

But presently a new noise was heard – the nightmarish disharmony of bugles and trumpets. And it was this noise which caused experienced 'Digger' platoon commanders like Len Montgomerie, Fred Gardner and Johnny Ward to pass the word round their sections: 'This is it, you blokes – sit tight and don't fire until you get the order.'

The ferocity in attack of the 3rd Royal Australian Regiment had won the admiration of everyone in Korea. But they had not, as yet, been severely tested in a defensive role. Brigadier Aubrey

Coad, always one of their more enthusiastic admirers, had confessed to certain misgivings about their performance in defence; after Kapyong he had no more cause to.

For the Australians yielded not one foot of ground and inflicted astronomical casualties on the Chinese, who tried to force a way through using the only advantage at their disposal – a massive preponderance of manpower. The 3rd RAR held their ground because they had men like Lieutenant Len Montgomerie and his No 4 Platoon of 'B' Company, who killed eighty-one enemy, mostly with the bayonet; like Corporal 'Evil-eye' Everleigh, who dispatched six Chinese in as many minutes, and Corporal 'Dogpatch' Davie, who bayoneted three and threatened to put a private on a charge for taking a prisoner; like Corporal Reg Parry of 'B' Company, whose section smashed three determined Chinese attacks within twenty minutes and accounted for twenty-three dead; like Corporal Bill Rowlinson and Private Dick Smith of 'D', both badly wounded and both adamantly reiterating in the most colourful language their refusal to be evacuated. Finally, like Lieutenant-Colonel Ian Fergusson, who had inherited this magnificent fighting battalion from Charles Green, and was fiercely determined that its record would continue to be a glorious one. Never for a single moment did his men disappoint him.

No two Americans did more to cement a strong bond of respect between the fighting men of two countries than two indomitable and unquenchably cheerful young tankmen of the 72nd United States Heavy Tank Battalion, Lieutenants Kenneth W. Koch and Wilfred D. Miller: time and time again they launched their tanks into counter-attacks against the milling hordes of Chinese, and times without number they braved withering fire to take up ammunition and evacuate casualties.

To make it a truly Anzac occasion, the 16th New Zealand Field Regiment had managed to extricate themselves from their beleaguered forward position with the South Koreans, and got back in time to pound the attacking Chinese with shells. 'Without those Kiwi gunners,' declared 'Evil-eye' Everleigh, who normally disliked New Zealanders on principle, 'we'd have been right in the flamin' cart.'

Away to the Australians' left, the 2nd Princess Patricia's Light Infantry were undergoing similar treatment and resisting with the same tenacity. To the inevitable accompaniment of bugles and supported by heavy mortar fire and hails of red tracer bullets, the Chinese launched five separate all-out attacks on the Canadian position; they found themselves pitted against men of the same calibre as those of 3rd R A R.

Two hundred screaming Chinese, stumbling over their own dead and wounded, launched themselves against Lieutenant Mike Levy's No 10 Platoon; according to an official account, Levy and his men 'literally disappeared in a swarm of Chinese': the expression 'wiped out' is frequently used in military history, thus implying the killing and wounding of every single man in the unit. It rarely worked out exactly like that, but it did in the case of No 10 Platoon: fighting back to back with bayonets, clubbed rifles, boots and fists, Levy and his men killed thirty before they went under.

A grenade fell into a trench occupied by Corporal Sam Douglas and three privates. Douglas, shouting to his men to jump clear, picked up the grenade and hurled it out of the trench, losing his right hand at the wrist in the process. In spite of this, he continued to command his section until he collapsed from loss of blood.

In the small hours of April 25th Captain Jim Mills' 'D' Company was attacked in such overwhelmingly large numbers that for the rest of the night the fate of the Patricias' position hung in the balance. The rest of the battalion was surrounded, and large numbers of Chinese had broken into 'D' Company's perimeter where furious hand-to-hand fighting was in progress.

A desperate situation called for desperate measures, and when Mills asked urgently for defensive artillery fire, giving a map reference, a shocked New Zealand voice said over the wireless: 'Jesus, sport, that's on your own position.'

Brodie's 29 Brigade had irreparably blunted the Chinese offensive; the Australians and Canadians exhausted and demoralized it; the 1st Middlesex and the 72nd United States Tank Battalion stopped it cold.

[3]

In June 1951 it seemed that the Chinese Communist Army was ripe for destruction: in one year of war the North Korean Army had suffered an estimated 600,000 casualties, including 100,000 prisoners; in only eight months the Chinese had lost another half million. After the fearful pounding that they had taken in the April offensive, the Chinese were having second thoughts about casually expendable manpower, and the poorly supported Communist army – Russia had let them down badly over supplies of tanks, artillery and aircraft – had demonstrated beyond all doubt its inability to sustain a major offensive for more than a few days.

General Ridgway, prior to his departure for Japan to take over as Supreme Commander in succession to MacArthur, had revitalized the 8th Army to such an extent that it was considered to be one of the finest fighting forces ever to be put in the field by the United States. Ridgway had made it so, and Lieutenant-General James Van Fleet kept it that way.

Van Fleet had graduated from West Point in 1915, from the 'Class of Generals', which had included Omar Bradley and Dwight Eisenhower. He had received his baptism of fire in the First World War, and as a machine-gun captain had been wounded just before the Armistice. But between the wars, like many another illustrious soldier bogged down as captain or major for years – in the British Army eighteen years as a subaltern was by no means unusual, and the record was believed to be twenty-one – apart from a glittering career as a footballer, no shining future was foreseen for Van Fleet. After 1944, however, when he led a regiment of the 4th Infantry Division on to Omaha Beach on D-Day, his promotion was rapid. In Greece he had given post-war Communism one serious setback, and now he was in Korea with the avowed intention of administering another.

The individual American soldier had lost much of his respect for the Chinese as a fighting man, and the word 'horde' was falling into derisive disuse: one of the jokes going round the front in the summer of 1951 was, 'I was attacked by two hordes and killed them both'. Now it was reckoned that three 'swarms' went to one 'horde', two 'hordes' to a 'human flood', and three 'human floods' to an 'inexhaustible reservoir of Chinese manpower'. It was a crazy assessment, but still good for morale.

Morale in the United Nations Forces was heightened still more when soldiers of all nationalities became eligible for 'R and R' – 'Rest and Rehabilitation' (but more universally known as 'Rack and Ruin') in Japan: the majority of soldiers felt the need of a period of rest after being in the front line.

Grimy and often verminous men, straight from bunkers, fox-holes, and pup tents, sped down the road in trucks with the inevitable Negro at the wheel; they always drove at breakneck speed, but the drivers of the 'R and R' vehicles rose to new heights of reckless daring.

Southwards through the remains of Uijongbu, through the shattered city of Seoul, over a bridge spanning the Han River and thence to the airfield at Kimpo – Americans, Britons, Australians, Canadians, Turks, Filipinos, Frenchmen, Belgians, Greeks, sped to Japan, all intent on spending their accumulated pay in the shortest possible time. Even the lowliest British private, the world's most impecunious fighting man, could expect to draw close on fifty pounds, and most of them spent it all in the five days: the record for immediate dissipation of funds was supposedly held by Australian Private Bob Simes, who ran through £130 and confessed that he remembered nothing about the first two days at all.

At the Commonwealth leave centre at Ebisu the men were greeted by avuncular sergeants, who informed them that the only parades were for meals and pay (neither of which were compulsory); they went to an armoury where they were relieved of their arms and ammunition; they were relieved of their clothes which were burned by disdainful hygiene sections, and of the Korean dirt under jets of hot water from showers that worked. Then they drew their pay and took the town for all it was worth. Once in Tokyo they were speedily relieved of their pay also.

And then with horrible suddenness the five days were up: they were broke again and it was back to the mud, the shells, the bullets and the smell of ordure and death . . .

[4]

Americans had long ceased to bleat 'How long have we got to go this war alone?' Gloucesters, Greeks, Belgians, Northumberland Fusiliers, Turks, Scots, Ulstermen, Frenchmen, Australians, and Canadians had taken care of that. But over vast tracts of the United States – in areas where the people, presumably, had never heard of Malaya or Africa – it was still felt that the British Commonwealth was not adding sufficient weight to the Korean War, which by this time had cost close on 15,000 American lives.

Some of this audible discontent percolated over the border into Canada, and Canadian action came promptly with the appearance in Korea of the 2nd Battalion Princess Patricia's Light Infantry. The month of May 1951 saw big additions to the Canadian contingent: in November 1950 a force designated 25th Canadian Infantry Brigade Group, had assembled at Fort Lewis, a United States Army camp near Seattle. Training at Fort Lewis was carried out at an exceptionally high (and, according to some Canadian soldiers) near murderous tempo under the tireless direction of Brigadier John Rockingham, who could march fifty miles, fire a rifle and thrust with a bayonet as competently as the toughest private in his command. 'Rocky' was something of a legend in the Canadian Army, and was later to become one in Korea: as a battalion commander, and later as an unusually youthful brigadier, he had fought in almost every major action in North-West Europe in the Second World War and had two DSOs to show for it.

During the first week in May it was soon apparent that the 25th Canadian was something of a 'Rolls-Royce' brigade, the term first wryly used by Aubrey Coad to describe the British 29th: it consisted of two infantry battalions with fine fighting records – the 2nd Battalion The Royal Canadian Regiment and the 2nd Battalion Royal Vingt-Deuxième Regiment, who were French and fiercely proud of it, and steadfastly refused to speak English. The brigade had its own tanks, 'C' Squadron, Lord Strathcona's Horse; the 2nd Field Regiment Royal Canadian Horse Artillery; the 57th Independent Field Squadron, Royal

Canadian Engineers and the 25th Canadian Field Ambulance. They even had a Field Punishment Unit: all possible aspects of Canadian military endeavour, and even lack of it, were catered for.

The 'Patricias', after Kapyong in the strongest possible position to tell the newcomers what war in Korea was like, were integrated into the 25th Canadian Brigade.

The 'founder members' of the Commonwealth Division – the 1st Argyll and Sutherland Highlanders and the 1st Middlesex – had returned to Hong Kong, and were relieved by the 1st Kings Own Scottish Borderers (whom we last met on a fractious drill parade in Kowloon) and the 1st King's Shropshire Light Infantry respectively. Both these battalions had been represented by volunteers in the earlier fighting: a number of Shropshires had been killed and wounded with the Middlesex, and the casualty lists of the Argylls contained the names of many KOSBs, most of them sustained on Hill 282.

Brigadier Aubrey Coad, who had drawn one of the toughest assignments ever handed out to a British commander in war, had left for a new appointment in Germany with promotion to Major-General and the CBE, an American Silver Star and the Order of the Legion of Merit (USA) to add to his two DSOs. In his place had come Brigadier George Taylor, also a double DSO, and an Army, Lancashire and Barbarians rugby football player: they liked them tough in Korea, and George Taylor had the right qualifications. The title 27 Brigade disappeared, and Taylor's new command became the 28 British Commonwealth Infantry Brigade composed of the 1st KOSB, the 1st KSLI, and those evergreen fighting men of the 3rd Royal Australian Regiment.

Also still in Korea was Brigadier Tom Brodie's battered 29 Brigade, now up to strength again. The 8th Hussars, adding tone to what otherwise might have degenerated into a vulgar brawl, remained, as did the Belgian Battalion, who unlike the Canadian 'Van-Dooze', strove to master the English tongue with often hilarious results. Those prime architects of destruction of Chinese, the 16th Field Regiment Royal New Zealand Artillery, the 45th Field Regiment and the 170th Mortar Battery stood to their guns, ready to give the enemy many more salutary hammerings. The 60th Indian Field Ambulance were on hand to dispense bandages to the wounded, medicine to the sick and potent curry to the hungry.

But this was considerably more than just a big build-up of Commonwealth strength. In earlier days the British, Australian and Canadian soldiers had been in Korea to be used at the whim of an American corps commander – large enough to be a reliable stop-gap in time of emergency, small enough to be shelved during a major offensive: an unsatisfactory situation, as Brigadiers Coad and Brodie had discovered on more than one occasion.

This was to be something quite new: for the first time in history troops from Britain, Australia, Canada, India and New Zealand were to be part of a Commonwealth operational division under a unified command. To command such a force clearly required a very exceptional soldier. Great Britain had one in Major-General James Cassels.

He was (and is) universally known as 'Gentleman Jim', not only because of his undeniably patrician good looks: Jim Cassels will preside at a policy meeting, congratulate a private soldier for an act of bravery, bust a sergeant-major, upbraid a sloppy staff officer, sack a brigadier, drink beer in a Sergeants' Mess, 'bowler hat' a major-general, cut a politician or journalist down to size, and listen to a complaint from a lance-corporal's wife – all with the same unfailing Edwardian courtesy and charm.

Jim Cassels also knew all about the 'sharp-end' of war: as a youthful subaltern of the Seaforth Highlanders, dressed in kilt and pith helmet with the regimental badge of stag's antlers prominently displayed thereon, he had first heard shots fired in anger on the North-West Frontier of India. At the age of 37 he commanded a brigade in North-West Europe and at 38 the 51st Highland Division; in common with many other Second World War generals, he was usually to be found considerably farther forward than was strictly necessary: like many another young officer in the period shortly after the Great War, Jim Cassels knew something about generals who, like the Duke of Plaza-Toro, led their regiments from behind; while their soldiers died, these doughty old gentlemen swilled claret in well-appointed *châteaux* and died in bed with a backwash of 20,000 good dinners and a thousand bottles of port behind them.

FOURTEEN

At the end of July 1951 the United Nations Line ran from a point on the west coast about twenty miles south of the 38th Parallel, north-east to the east coast approximately 20 miles north of the Parallel. It cut the Parallel some 40 miles north-east of Seoul.

It was a stable line and, so thought General James Van Fleet, a virtually impregnable one: although there was no close contact with the Chinese at the time, it was clear that their recent offensives had been brought to a complete standstill, if they had not petered out altogether.

Communist Korea had become an impotent shambles: its railroads had been bombed out of existence, its communications were crippled, its industry, never brilliantly productive, was now totally ineffective. All Communist supplies had to come from the Soviet Union, and Russia had on several occasions proved to be a broken reed.

Van Fleet, like Douglas MacArthur, was imbued with the offensive spirit; to him it was unthinkable that there could be any 'substitute for victory'. MacArthur had hoped to win the war at a single stroke at Inchon; now Van Fleet had his own blueprint for victory, but it would almost certainly be horribly expensive in American lives: a series of amphibious landings on the east coast, simultaneously with a break-out drive to the north-east.

'At that time,' said Van Fleet, 'the east coast did not have a big build-up of defensive forces, and we could easily have made landings there. The Navy could have shot us ashore and kept us ashore as we built up. We had the Chinese whipped . . . they were in awful shape . . . they had no conception of fast moves . . . they had no system of communication.'

Van Fleet resembled a boxer who, after landing a fusillade of crippling body blows, has opened up a deep gash from which his opponent's life blood is pouring all over the ring. The United Nations were poised for a knock-out blow: he had Marines for the assault landings and a revitalized Army for the break-out, not forgetting a Commonwealth Division.

What could possibly stop him?

The answer was the United States of America. Sick of a war, which had already caused 80,000 casualties, the United States was ready to grab any opportunity to bring this unpopular conflict to an end. Such an opportunity seemed to be at hand, for a whisper was going round that the Chinese might be ready to sue for peace. While this was heavily discounted in most military circles, it represented a straw to clutch and America reached out for it hopefully.

The whisper became a roar, but initially it seemed to have no conceivable connexion with peace – quite the reverse: in a radio broadcast Mr Jacob Malik, Soviet delegate to the United Nations, devoted fifteen minutes to a noisy and blistering denouncement of the United States and NATO. Running out of breath, telling phrases and invective, and running overtime as well, Malik eventually got around to the subject of peace.

'The Soviet peoples,' said Malik in a slightly more muted key, 'believe that the most acute problem of the present day – the problem of the armed conflict in Korea – could also be settled. The Soviet peoples believe that as a first step discussions should be started between the belligerents for a cease-fire and an armistice providing for the mutual withdrawal of forces from the 38th Parallel.'

After this broadcast the United States Ambassador in Moscow, Mr Alan Kirk, was instructed to obtain a clarification of the Malik statement: in doing so, Kirk's powers of diplomacy were tested to the utmost.

This resulted in General Ridgway being authorized by the United Nations Assembly to invite the enemy to discuss in the field the possibility of arranging a cease-fire. It also resulted in the shelving of General Van Fleet's plan for victory in Korea.

So, at Kaesong on July 10th, 1951, the cynical and ill-tempered bickering and squabbling, variously referred to as 'peace talks', 'truce talks' and 'cease-fire talks' started. They were to continue intermittently for the next two years.

Meanwhile, men continued to die in Korea . . .

[2]

The acrimonious haggling, which went by the name of 'peace talks', proceeded on its quarrelsome way. Admiral C. Turner Joy and his team of high-ranking American sailors, soldiers and marines strove to inject some sense into the meetings, which were obstructed at every turn by the Chinese delegates who were petulant, insulting, sarcastic and deliberately obtuse by turns: there was General Nam Il, who produced guffaws of laughter for no attributable reason, chain-smoked and hoicked loudly at every session; there was General Lee Sang Cho, who was so anxious to impress the American imperialists with his iron self-control that he permitted flies to crawl over his face.

As a result of the peace talks, General Van Fleet was authorized by Ridgway to carry out only 'limited offensives'. Ridgway, a hundred per cent offensively minded soldier himself, sympathized with Van Fleet when his plan for a swift victory was turned down. Of the shelving of Van Fleet's plan he wrote: 'Military men and statesmen, too, will long debate the wisdom of stopping that proud Army in its tracks at the first whisper that the Reds might be ready to sue for peace. To my mind it is fruitless to speculate on what might have been. If we had been ordered to advance to the Yalu we could have done it – if our Government had been willing to pay the price in dead and wounded that action would have cost.'

The United States were not willing to pay this price, and the Static War began.

[3]

For the soldiers in the front line there were the Hills of Korea: the cratered and shell torn, rat and lice infested peaks which the Americans called 'real estate'. Like thousands of other hills in two world wars – in Africa, Italy, North-West Europe and

Burma: hills had to be clawed up by painful inches in the face of lashing fire and hails of hand grenades; to be captured and held; to be endlessly and back-breakingly dug and wired, for work on defensive positions was never finished; to be shelled off, overrun and re-taken; to be buried alive on or to die on – hills with names like Pork Chop Hill, the Punch Bowl, Frostbite Ridge, Old Baldy, Bloody Ridge and The Brown Bastard; hills which became 'home' for the soldiers who fought for them.

In November 1951 each battalion of the Commonwealth Division had its own hill in the new Jamestown Line, formed after the October advance. Baulked of his plan to carry out amphibious landings on the east coast and to drive forward up to the Yalu, General Van Fleet had been permitted by Ridgway to carry out only a 'limited offensive' of 8,000 yards to a more commanding position which, it was hoped, would lend more weight to the truce negotiations: in their present state of bogged down and bickering stalemate, they needed all the weight they could get.

On one of these hills was the 1st Battalion The King's Own Scottish Borderers: they had had to fight hard for their tenancy, as had the 1st Leicesters (who had relieved the Northumberland Fusiliers), the 1st KSLI and the 3rd Royal Australian Regiment; the Canadians in their advance had experienced a comparative walkover.

Now Hill 355 was the KOSB's home: they had fought for it and thirty-one of them had suffered death or wounds for it. If the Chinese thought that they could shift them off it, they were very welcome to have a try.

The Chinese were fiercely determined to remove the KOSB from this hill, and the Scots were equally resolved to stay on it.

The morning of Sunday November 4th started quietly enough with the by now conventional shelling of forward positions to which every man of the Commonwealth Division was by this time acclimatized.

Chinese artillery fire, however, was still something of a novelty, and Brigadier William Pike, General Cassel's chief gunner, maintained that the Chinese were using the peace talks advantageously to build up military strength. To add subtle irony to the situation, while the sound of battle waxed and waned in the dark valley between the British Commonwealth and Chinese lines, a plainly visible finger of light marked the site of the new 'Peace Village' at Panmunjom: the talks at Kaesong had been a conspicuous failure so far, and a new location had been chosen.

Chinese artillery fire was definitely hotting up, and the men of the KOSB, KSLI, and the 3rd Royal Australian Regiment became increasingly aware of this as the morning of November 4th gave way to the afternoon, and casualties started to mount. The enemy guns were particularly difficult to locate and destroy, as they were cleverly concealed in tunnels in the hills opposite the Commonwealth Divisional front: while General Nam Il laughed, prevaricated, smoked and expectorated out of (and frequently inside) the peace-talk tent, the Chinese engineers had not been wasting their time.

Brigadier Pike, however, was not unduly perturbed at this stage: he had the means to distribute the fire of the mass of guns at his disposal over every point in the Chinese defences in his sector and on their likely lines of attack. The Royal Regiment of Artillery had never failed the infantry yet in Korea, and they were not going to now: in the next thirty-six hours Pike's gunners were to fire the staggering total of 30,829 rounds.

[4]

The bombardment raved louder – louder than the briefest thunderstorm that any of the men of 28 Commonwealth Brigade had ever heard; louder than an avalanche or the explosion of volcanoes. A monstrous tearing roar filled the earth and the sky, as an absolute fury fell over the KOSB's position, in earshot of the stalling Chinese and, by this time, fermenting American peace delegates. The ridge held by the three forward companies, 'B', 'C', and 'D' had, in the space of ten minutes, become an inferno of smoke, dust, flame, and explosions. None of the company commanders – Majors Harrison, Little, and Robertson-Macleod – were strangers to war and shelling, but all three agreed that any pounding which they had experienced in the Second World War paled into insignificance compared to this maelstrom. Of the 10,000 shells which fell on the Commonwealth Division on November 4th, it was calculated that 6,000 of them fell on the Borderers.

In the Battalion command post, which bore a marked resemblance to a set from the play *Journey's End*, were Lieutenant-Colonel Dennis Tadman and his Adjutant, Major Alan Jackson: Tadman had succeeded to the command of the Battalion on the promotion of Colonel John MacDonald to Brigade Commander of 28 Brigade. Partially deafened, as was everyone else, Tadman knew what would happen when the barrage lifted: the Chinese would launch a full-scale infantry attack at men whom they thought must be either dead, wounded, buried alive or too shocked and dazed to resist. If the shelling continued at its present rate, thought Tadman grimly, they very likely would be.

The Borderers could do nothing but stay in their trenches and take it – gritting their teeth, impotently swearing, chain smoking (the consumption of cigarettes in the KOSB on that day was prodigious), trying to burrow ever deeper into their holes. Meanwhile, every available means of defensive fire – aircraft, artillery, and mortars – were called upon. The KOSBs drew slender comfort from the knowledge that the Chinese were getting almost as heavy a hammering as they were.

In a trench on 'B' Company's position reposed the bulk of a private soldier, so huge that he seemed to be overflowing out of it. This was Private 'Big Bill' Speakman: just 24 years old, measuring six feet and six inches from the crown of his bullet head to the soles of his size 12 boots, and weighing fifteen and a half stone, of which only a pound or two were superfluous (he has put on a stone since then, but no one will deny his entitlement to it). If some space is taken in a description of him, it is because he is a subject worthy of it.

Bill Speakman is not a Scot; indeed, originally he was not even a King's Own Scottish Borderer – in the closing stages of the Second World War, he enlisted in the Black Watch, and there were many 42nd Highlanders, not fools or romantics, who were ready to swear that his appearance on the scene earlier would have appreciably shortened the war.

He was born and bred, in fact, in Altrincham, Cheshire: a slow-moving, slow-speaking giant, but an innately gentle one, like Garth of the *Daily Mirror* strip cartoon.

Bill Speakman had a conduct sheet of impressive dimensions when he first joined the 1st KOSB, but his military 'crimes', if they can be so described, were usually of omission rather than commission: he had but the vaguest idea of time, and frequently failed to read company detail with the result that the guard was often mounted without Private Speakman, W.; temporarily incapacitated by some wildly celebratory occasion, he had more than once spent the night in ditches and other unlawful places of assembly on his way back to barracks. He lost items of kit at regular intervals which, because of his size, were well-nigh impossible to replace, and was reckoned to be one of the most impecunious soldiers in the 1st Battalion The King's Own Scottish Borderers.

And now Bill Speakman was on a hill in Korea: he was being viciously shelled, which he disliked; it seemed that he and everyone else were forced into permanent defensive, and this he disliked even more. Like most very big men, he was slow to anger but terrible in his wrath; as the shells rained down on the KOSB, 'Big Bill' was just coming to the boil . . .

[5]

Colonel Tadman had thought that the Chinese attack would come in immediately after the shelling ceased: in this surmise he was not quite correct, for the long lines of Chinese in their olive-green quilted uniforms advanced shoulder to shoulder into the barrage of not only their own shells but those of Brigadier Pike's guns as well. According to the men on the spot, it would have made no difference if shells had been coming from both flanks as well – the Chinese would still have come on. Clearly the Chinese high command had reverted once more to its former haphazard non-reportage of casualties.

In front of the KOSB's position there had been a positive

honeycomb of barbed wire, the result of hours of back-breaking toil. But the front might just as well have been protected by strands of cotton: most of it was blasted away by friendly and enemy shells alike, and the Chinese brought up bangalore torpedoes and pole charges to blast a way through what little was left. If any fences still remained standing, they forced a way through by a method taught at the more enterprising battle schools in Great Britain during the war – the first man to come to the wire flung himself face downwards upon it, while his comrades used him as a human stepping stone. In this manner some 6,000 Chinese came directly into the sights of the machine guns, mortars and rifles of the King's Own Scottish Borderers.

The attack was carried out to the usual raucous accompaniment of bugles and horns, and a new propaganda device, which would have caused the KOSBs to laugh if they had been able to snatch a moment's respite from the serious business of killing, had been brought into play: a blaring voice, identified by a wag of 'B' Company as belonging to Doctor Fu Manchu himself, was heard over a loudspeaker, and he had a message of some urgency to deliver. 'Scotsmen go home!' exhorted the voice; 'go home or this hill will become your grave!'

It did, in fact, become the grave of seven of them in addition to 87 wounded and 44 missing, but they inflicted casualties on the Chinese which were enormous by any standards. In this battle Chinese fanaticism was more than matched by Lowland Scottish guts.

Second-Lieutenant Hugh Paterson, commanding one of the forward and most exposed platoons coined his own phrase to describe the attacking Chinese: 'Like flies on the back of a mule'. Paterson and several other men, who could almost see the whites of the attackers' eyes as they pumped bullets into them, swore that they were drugged.

It seemed to be the only logical explanation for the fantastic behaviour of one Chinese light machine gunner, who sneaked up a crawl trench to within five yards of Paterson's position. He set the machine gun up, wearing a broad smile and crooning quietly to himself. Paterson fired at him and the bullet hit the gun's magazine before ricocheting away in a flurry of sparks.

Apparently totally unconcerned, the Chinaman continued to set the gun up, still wearing the same vague and disengaged grin; he was still smiling when Hugh Paterson put the next round through his brain . . .

The Manual of Small Arms Training laid down that for maximum effect the light machine gun should be fired in bursts of three or four rounds. Any weapon-training expert who expounds this theory to KOSB bren gunners Lance-Corporal Morris Duff, and Privates Robert Bryson, John Pender, David Carstairs, and 'Chick' Crabb will get a short and almost certainly unprintable answer; having got this off their chests, they will say that the only possible way to engage an enemy like this was to press the trigger and keep it pressed in the pious hope that the 'Number Two' on the gun had a fresh magazine instantly to hand.

Private Ken Johnson, a Kentish fish out of water in a Scottish regiment, had performed on the rifle ranges without any sort of distinction in Hong Kong: on more than one occasion Company Sergeant-Major 'Busty' Murdoch, on inspecting Johnson's target, had been moved to tear his hair and groan out loud. 'But on that day,' declares Johnson (now a crack shot and a company sergeant-major himself), 'even I couldn't miss. Any bloke who could hit a haystack at ten yards could have been sure of killing at least five Chinamen in that battle.'

The battalion's mortars, commanded by Lieutenant Peter Rooke, fired with deadly effect and discharged some 5,000 bombs in four hours. But one does not need to be a ballistics expert to know that mortars will not fire indefinitely without water and every man of the mortar platoon had emptied his water bottle into the barrels already. Many were the calls for mortar support from 'B', 'C', and 'D' Companies, and Peter Rooke answered them all unstintingly. But towards nightfall on November 4th, Rooke had to tell Major Philip Harrison, commanding 'B' Company, that further mortar support was impossible: the barrels were smoking and red hot, and there was no water.

'Pour beer down 'em,' said Harrison shortly; he had been hit in the back by a potato-masher grenade, and buried alive: to this day he is not sure whether it was as a result of an enemy shell or a friendly one, for like company commanders of the Gloucesters and 5th Fusiliers on the Imjin, he had had to call for defensive artillery fire on his own position. If his answer seemed inclined to a certain facetious curtness, he may be forgiven.

Beer . . . forty-eight bottles of this precious liquid lay in the mortar-platoon position, as yet untouched by human hand or enemy fire: forty-eight pints of cold amber liquid, two per man, for which any member of the platoon would willingly have bartered his soul.

Rooke indicated it to the platoon sergeant, Sergeant Bill Reeve, and gave a single crisp order: 'Pour it into the mortars.'

'Pour it into the mortars?' echoed Private Walter Nelson in tones of shocked incredulity; 'but that's our beer ration!'

'You can count yourself lucky if you ever get another, son,' said Reeve grimly. 'Now, get them bottles open and look sharp about it . . .'

The mortar platoon of the 1st KOSB was in business again . . .

[6]

The Distinguished Service Order has only once been awarded to a National Service officer; on November 4th, 1951, 19-year-old Second-Lieutenant William Purves made history.

Before his call-up it had seemed unlikely that he ever would. He occupied a junior position in a Kelso Bank, and if his rather hesitant and diffident air was anything to go by, his advancement seemed likely to be slow. Neither did Colonel John MacDonald, who knew a lot about soldiers but hardly anything about junior bank officials, foresee an illustrious military future for him. Neither, to be honest, did William Purves himself: he was in the Army because the Government said he had to be; he had satisfied the comparatively unexacting requirements of a War Office Selection Board as to his suitability to hold His Majesty's commission. Then had followed Eaton Hall, forcing-house of conscript officer cadets, a brief period at the Regimental Depot and then Hong Kong. Like hundreds of other young National Service second-lieutenants, Purves credited himself with few qualities of leadership and doubted if they would ever be put to the test.

Colonel Dennis Tadman had plenty to worry him as night fell on November 4th, and the main cause for his anxiety was the fate of two platoons – one of 'C' Company and one of 'D' – on an isolated position known as 'The Knoll': the 'D' Company platoon, commanded by Second-Lieutenant Barney Henderson, was in the most forward position; slightly rearwards, on a ridge which because of its peculiar shape was known as 'Lovers' Knot', was the 'C' Company platoon under Second-Lieutenant William Purves. Every officer and man of the 1st KOSB was fighting desperately by this time, and every position was in grave danger of being overrun, but 'C' Company and 'D' Company men were in a special category of peril.

There had been no news from this force since the battle started. But towards 2 AM a message got through: they were completely cut off; they were being attacked from three sides; there were thirteen wounded, including Second-Lieutenant Henderson. What were the orders?

The voice was that of William Purves. Tadman's first instinct was to hope that it was Henderson in command, because he had the ingrained pre-war regular soldier's preference for the Sandhurst-trained professional rather than a junior bank clerk turned soldier. But he soon dismissed this out-of-date thought: one must move with the times, and Purves sounded steady and confident enough.

What Colonel Tadman did not know was that Purves had been shot through the shoulder just one hour earlier; there had been no time to dress the wound, and his battledress blouse was soaked with blood from the upper arm to the wrist.

Neither did he know that for the past five hours Purves had been moving about in the open under withering fire, encouraging and inspiring the men under his command and hurling grenades at the Chinese swarming round his position. The spirit of the leader very quickly became the spirit of the men, as an encouraging pile of dead round the beleaguered outpost testified.

The 1st KOSB were in a predicament not unlike the one in which the Gloucesters found themselves on the Imjin battle:

'B', 'C', and 'D' Companies were all fighting magnificently, but casualties were mounting steadily, ammunition was running out, wireless sets were being rendered unserviceable one by one. Colonel Tadman decided to make a limited withdrawal and to tighten the perimeter.

'Get out of it,' he ordered Purves.

'We're holding them, sir.' There was a note of pride and perhaps a little of protest in Purves' voice.

'You heard the order. Get back here.' The line went dead, and Dennis Tadman made a mental resolve to think more kindly of National Service subalterns in the future.

Just before midnight Purves, his wound still undressed, gave out his orders to the forty men of the two platoons still on their feet: they were going to fight their way back to battalion headquarters; every wounded man would be carried out, as would all arms and ammunition. Were there any questions?

At about 3 AM on the morning of November 5th, as the battle continued to rage unabated outside, Second-Lieutenant William Purves staggered into the log-bunker battalion command post. His left arm, the sleeve of his tunic saturated with blood, hung limply at his side; with his right he executed a salute of exemplary perfection and reported: 'Two platoons of "C" and "D" Companies ready for action, sir. What are the orders?'

The orders for Second-Lieutenant Purves were to proceed immediately to the Regimental Aid Post, where Captain George Ruttledge, the Medical Officer, and his staff were working ceaselessly and tirelessly, dealing with the flood of casualties from the forward companies; the orders for the rest were to go to their companies and continue to kill Chinese with as much strength as God had given them . . .

And that is how William Purves, the 19-year-old bank clerk from Kelso, became the first National Serviceman to be admitted a Companion of the Distinguished Service Order . . .

[7]

In 'B' Company Headquarters dug-out it was plain that the customary bland good humour of Private William Speakman, that amiable giant, was fast deserting him. Speakman was company runner and, by the same token, company commander's bodyguard. But there was no running to be done, and Speakman considered that Major Philip Harrison had already proved beyond all doubt that he could manage very well without a bodyguard.

'Big Bill' was in a temper, and not only because the beer was finished: Company Sergeant-Major 'Busty' Murdoch was a firm believer in the adage 'rank hath its privilege', and had seen to it that Company Headquarters was well supplied. During brief lulls in the fighting, appreciative gurgling noises from the throats of Sergeant Jock Duncan, Corporal Tug Wilson, Corporal Eric Wood, Lance-Corporal Ed Buchanan, and Private Bill Speakman – the loudest and most appreciative noises of all came from the two last named – told the CSM that his foresight was appreciated.

What principally annoyed Speakman was that apparently the Chinese were occupying a position which rightfully belonged to 'B' Company: Nos 3 and 4 Platoons, fighting every inch of the way, had been forced back towards Company Headquarters and from all accounts No 5, fighting equally hard but similarly depleted by casualties, was in a precarious position. Bill Speakman did not like this, and he was not going to have it; he decided on his own initiative, a quality which he had displayed but rarely in the past, to drive the enemy off the position and keep them off it.

Easing his large bulk to a sitting position, Speakman calmly and methodically proceeded to load himself with hand grenades – he stuffed them into his pockets, his ammunition pouches and down the front of his shirt. He then left the dug-out.

'And where the hell d'you think you're going?' demanded CSM Murdoch.

'Going to shift some of them bloody Chinks,' replied Speakman, a man of few words, and was gone.

Speakman charged alone at the Chinese on the ridge and hurled his grenades, a round dozen in all, into their midst. Anguished howls in the half-light told him that his aim had been good. He then made his unhurried journey back to Company

Headquarters, and met Major Harrison on the way. And Harrison, too, asked him: 'Where are you going, Speakman?'

'Just going to pick up some stores, sir,' said 'Big Bill': later Harrison recorded that he met Speakman on no fewer than ten similar errands.

Back in the dug-out, 'Big Bill' collected a further load of grenades. To the other occupants he said: 'Knocked off a few of 'em. Who's coming?'

The answer was immediate and shattering: the CSM's proper place was at Company Headquarters, but Major Harrison was in the thick of the fight somewhere else, as he had been since the battle started; Sergeant-Major Murdoch had put Private Speakman on a charge more than once before for slowness or plain idleness in carrying out orders, but there was something infectious about his purposeful belligerence. The CSM, Duncan, Wood, Buchanan and Wilson followed him out of the dug-out with alacrity.

Speakman led the party in a series of grenade charges against the Chinese, and continued to do so as each successive wave of the enemy reached the crest of the hill. The force and determination of these charges broke up each enemy onslaught and resulted in an ever-mounting pile of dead.

In all they made some fifteen charges, and every one of the party was wounded; Speakman, indeed, was wounded twice, but paused only long enough to have a field dressing applied to a gaping hole in his leg before he was back in among the enemy, his grenade-throwing aim totally unimpaired.

And then of a sudden the men of 3, 4 and 5 Platoons, who had been forced off the position, ceased to shoot from the trenches behind this gallant little band and a voice cried 'Bill Speakman!' It passed down the line, and, as one man, they came back howling 'Bill Speakman!' They drove the Chinese out and fell back into the lost trenches. As dawn broke on November 5th, they were still there, waiting for the next attack. Astonishingly, it never materialized.

Subsequent intelligence reports revealed that the King's Own Scottish Borderers had been attacked by a whole Chinese Division, some 6,000 strong and that, in the course of the battle, the enemy had casualties of over 1,000 killed. The Battalion's own casualties were not light: seven killed, 87 wounded and 44 missing.

In addition to the DSO won by William Purves, Lieutenant-Colonel Dennis Tadman, and Majors Harrison and Robertson-Macleod were also admitted to the Distinguished Service Order. There was a Distinguished Conduct Medal for 'Busty' Murdoch, and Military Medals for Lance-Corporal Buchanan and Corporal Wood.

For Private Bill Speakman there was only one possible award – the Victoria Cross.

PART FOUR

The Static War

JULY 1952 - JULY 1953

FIFTEEN

It was beginning to be a new sort of war altogether – a static war, reminiscent of the trench warfare of 1916; a war horribly intimate, uncomfortable and unorthodox; a war in which the infantry soldiers lived on perpetually close terms with their own gunners and the enemy's. Now gains on either side were reckoned in yards rather than leagues.

'Take a Christmas card valley,' wrote a war correspondent, who was running out of eye-catching headlines, 'add tents and sandbags, sprinkle well with stove pipes, place a latrine in the middle and you have a cosy corner in Korea.'

This was a corner of the Commonwealth Division in the grip of the 1951–2 winter: climatically the season was every bit as vile as the one that had gone before, but there was one essential difference: during the freezing misery of 1950–1, operations had been fluid, and the Argylls, Middlesex, and 3rd Royal Australian Regiment had been constantly on the move. The word accommodation was an echo of a more spacious past, and virtually did not exist: the men of Aubrey Coad's brigade spent most of their time marching and fighting, and on the few occasions when they were doing neither, they were in slit trenches or bunkers with the most meagre overhead protection against the elements. More precious than beer or cigarettes in the early days was wood with which to make a fire; it was infinitely more important than women, the mail or the possibility of a home posting.

But faced with the prospect of staying in the same position for a long time, and by the end of November a major offensive by either side seemed to be a remote possibility, the soldiers of the British .Commonwealth Division, displaying enormous ingenuity, set about the serious business of making themselves as comfortable as possible. Any fool can be uncomfortable, was the oft-quoted dictum, and a true one.

There was little to commend Korean houses as semi-permanent residences: few of them were undamaged, and one and all were infested with rats and lice. For reasons of health and self-preservation, the Commonwealth Division went underground as thoroughly as any moles.

Considering the climate, they achieved a degree of comfort, even in the most forward positions, which was nothing short of miraculous. They lived in dug-outs, revetted with local timber and well roofed. Oil stoves were available, and the Officers' Mess of the 1st Royal Norfolks – hitherto there had never been such a luxury as an Officers' Mess – even had an open fireplace. The Norfolks also had home-made armchairs and a garden swing which would have cost £40 at Harrods. As in the First World War, officers' messes had their murals: they had moved on a bit since the Kirchner girls, but the message was the same. They also had plentiful supplies of liquor, which was the constant envy of the Americans, who still fought the war in a mood of resigned temperance.

Everywhere along the Divisional front the domestic risk of fire, explosion and collapse was ever present. But it can be said that the soldiers achieved the impossible – they made a success of living in conditions which were the worst since the Crimea, and it was entirely due to their own ingenuity and powers of improvization. The more enterprising units even achieved showers with hot and cold water, and a Sergeants' Mess without a bar was considered to be downright sordid.

There was also 'R and R' leave in Japan to look forward to, and in 1952 the Commonwealth Division opened its own Rest Centre at a small village by the sea, near Inchon. Florence Nightingales had arrived in Korea at last, in the elegant shape of two indefatigable ladies of the Women's Voluntary Services, the Misses Hilda Wood and Patricia Whittall. At this centre, in an establishment imaginatively styled Ye Olde Crown Inn, the staple delicacy of the British soldier – ham, eggs, and chips – was available in unlimited quantities. The Misses Wood and Whittall occupied a place in the affections of Commonwealth soldiers to which Miss Diana Dors could only aspire in a limited sense.

Professional artistes began to put in an appearance in Korea:

for the Americans, Danny Kaye, Bob Hope, and Errol Flynn; for the British, Ted Ray, Jack Warner, Larry Adler, Carole Carr, and Frankie Howerd.

Brigadier Tom Brodie, prime architect of the Chinese defeat on the Imjin, had gone home, having earned the admiration and respect of everyone in Korea. He had been succeeded by Brigadier Abdy Ricketts, a Wykehamist and Durham Light Infantryman, who had all the right credentials for a tough campaign: he had first heard shots fired in anger as a pink-cheeked subaltern in Shanghai, and had graduated, via the North-West Frontier of India in its most turbulent period, to the command of a Chindit brigade in Burma.

The composition of 29 Brigade was now as follows: 1st Battalion the Royal Norfolk Regiment, 1st Battalion the Royal Leicestershire Regiment, and 1st Battalion the Welch Regiment.

The 8th Hussars, who had given magnificent tank support where it was needed most, had been relieved by the 5th Inniskilling Dragoon Guards – another Balaclava Regiment, and imbued with the same belligerent and carefree spirit of the '*arme blanche*': the 8th Hussars had complained that much of the country was untankable; the 5th Dragoon Guards would soon rectify that.

The 2nd Battalions of the 'Princess Pat's' and the 'Van-Dooze' were relieved by the 1st Battalions of the same regiments: they came to Korea knowing that they had tough traditions to live up to; the 3rd Royal Australian Regiment was relieved by the 1st Battalion the 3rd had become something of a legend in Korea, and the 1st were fiercely determined that they should fight as well. As if to make sure that they did, a familiar and battered face was seen in their ranks: by some complicated manipulation of postings, which could only have happened in the Australian Army, Corporal 'Evil-eye' Everleigh, complete with a Military Medal won in the early 1950 fighting, had reappeared in the ranks of the incoming 1st Battalion for a second tour in Korea. A 'Ratty Digger' indeed . . .

The appearance of the Black Watch in a peacetime station has sometimes caused a garrison commander to shudder apprehensively. But a field commander, in any theatre of war, immediately rubs his hands together in glee at the first intimation of their approach. When General Jim Cassels heard that the Black Watch were under orders for Korea, he was pleased enough about the arrangement: if he could not have the Seaforths, his own regiment, then the 42nd Highlanders would do well enough. His only feeling of relief was that they were not arriving together . . .

In any theatre of war, wherever the fighting is thickest and the climatic conditions most vile, Londoners have always been prominent: there had been the 1st Middlesex, a predominantly Cockney battalion in the early days; now the 1st Royal Fusiliers, the City of London's own regiment, were under orders for Korea. The Royal Northumberland Fusiliers had fought hard and consistently well, and yet another 'Geordie' regiment, the 1st Durham Light Infantry would soon be there too, as would the 1st Duke of Wellington's Regiment – the officer roll of the 'Dukes' reads like an international rugby programme at Twickenham: England's halves, Second-Lieutenants Denis Shuttleworth and Michael Hardy; Scottish internationals, Second-Lieutenants David Gilbert-Smith, and Michael Campbell-Lamerton.

Liverpool, a constant breeding ground of good fighting men, was not going to be left out, and the 1st King's (Liverpool) were also on their way to Korea.

[2]

The 'peace', 'truce', or 'cease-fire' talks – they changed in nomenclature almost hourly – continued on their acrimonious and unproductive way. From the start the United States and Chinese delegates had disliked one another on sight; by early 1952 they made little attempt to disguise the fact.

The Chinese were as truculent and obstructive as they had been ever since the talks started in July 1951; the Americans were becoming increasingly embarrassed, as it became apparent that the Chinese were merely using the talks to build up their military strength.

The talks went on for days, which lengthened into weeks, and for weeks, which lengthened into months. For the American delegates it was more difficult than arguing with a woman; infinitely more difficult than trying to read a newspaper in a high wind. On the vexed subjects of demarcation lines, exchange of prisoners, temporary cessations of hostilities and demilitarized zones, the same dreary deadlock was reached. It is small wonder that the American delegates had to be changed at frequent intervals: high-ranking soldiers, sailors, airmen, and marines all pitted their wits against the inscrutable Chinese, and all confessed to heightened blood pressure and a permanent hatred for oriental politics.

Meanwhile, men continued to die in Korea.

[3]

The press were bored for the best of all reasons: there was nothing to write about; in 1952 there was scarcely an action which earned more than six inches of newsprint in any British national daily. During 1950 and 1951, men like Frank Owen, Bernard Wicksteed and Cyril Aynseley had served it up hot and strong from the sharp end: the arrival of 27 Brigade, Hill 282, the Imjin River battle, the Australians and Canadians at Kapyong, the KOSB's November battle – these had all been naturals.

But there was nothing new to say about the Korean winter, and the peace talks defied intelligent journalistic analysis. British newspapermen, however, had a duty to perform to 'Mum' – the mothers of the teenage crusaders, who lived in their underground 'hoochies', stood-to at dusk and dawn, patrolled endlessly in no-man's-land, dug, wired and sometimes died. Waiting for a major offensive which never came, waiting for the weather to improve, waiting for some sensible outcome to the peace talks, the British pressmen turned to the universal journalist's standby in times of static and uneventful warfare – the 'local boy story'.

Arthur Helliwell of the *People*, one of the most widely travelled and indefatigable of them all, a by-line correspondent all round the world, took as his subject Private Sammy Wilson of the 1st KOSB – a perky Glaswegian with a perpetual gap-toothed grin, whose primary concerns were the performance of Queen of the

South in the Scottish Football League and the lack of thick, creamy rice pudding for dinner. From the front line Arthur Helliwell wrote an open letter to Mrs Wilson and to the Mum of every other National Serviceman in Korea, and did it with his usual professional competence.

Mrs Wilson had no cause to be worried about Sammy, wrote Helliwell, even though he had not had a bath, changed his clothes, removed his boots or seen a film show for two months. He ate and slept in a tiny, sack-lined bunker in which it was impossible to stand upright. But it had a stout, timber-reinforced roof which afforded protection from shells and mortar bombs. By Korean standards, Sammy Wilson considered it to be a desirable enough residence. All right so far, Mum . . . ?

Sammy asks for very little: a letter from home, a chance to read a month old newspaper, a fortuitous bar of chocolate or bottle of beer, a few days out of the front line, a pair of clean socks. 'I marvel at his cheerfulness,' continued Helliwell; 'Sammy just refuses to be downhearted, and he still wears a smile that would gladden your heart if you could see it. And don't forget that rice pudding for dinner when he comes home again.'

Sammy Wilson was one of the thousands of boys who had become men the hard way. The larger issues of the Korean War did not interest them a great deal; the peace talks had gone on for so long that they had lost all faith in them, as everyone was beginning to: they still had a war on their hands, and they had to fight it whether they liked it or not.

The Sammy Wilsons in Korea knew nothing and cared less about the peace talks, and the attitude was becoming infectious: the Chinese still, apparently, did not know what the word 'agenda' meant; interminable arguments went on about neutral zones (whatever they may have been); accusations of breaches of previous arguments were a daily occurrence. Whenever an agreement seemed remotely likely, the Communists speedily raised some new irrelevancy and adroitly sabotaged the discussions by some new accusation of bad faith.

In the spring and summer of 1952 celebrities of all nationalities, anxious to view this strange war at close quarters, appeared in Korea in questing droves – they were greeted with Guards of

Honour and eighteen gun salutes; they dined in the Commonwealth Division Officers' Mess; they travelled in helicopters – the Archbishop of Quebec and Danny Kaye; the Moderator of the Church of England in Canada, and Lady Louis Mountbatten; Field-Marshal Lord Alexander, Minister of Defence, and Mr Selwyn Lloyd, Minister of State; General Eisenhower and Jack Warner; Admiral Sir Rhoderick McGrigor, the First Sea Lord, and the Bishop of Korea; Mr Adlai Stevenson and Larry Adler; the Countess of Limerick and Ted Ray, the Bishop of Leeds, and Carole Carr. Everyone wanted to have a look at this war, which was like no other war in history.

[4]

Suddenly it was summer again: the weather changed from frozen misery to sultry heat and drenching downpours of rain – roads became churning nightmares of mud; bridges and bunkers collapsed; the Imjin River rose sharply, and there was anxiety for the safety of the bridges over the river by which the troops of the Commonwealth Division were supplied. In the event the river mercifully subsided, the bridges remained intact, and the Canadians played ice hockey.

Major-General Jim Cassels had gone home on promotion to the command of an Army Corps in Germany: 'Gentleman Jim' had seen the British Commonwealth Division grow from nothing into one of the finest fighting formations ever to take the field. It was largely due to him that the Division's morale remained sky-high in circumstances which would have excused a sharp decline.

His successor, Major-General 'Mike' West was cast in the same mould: like Cassels, he was a big man in every way and had commanded a brigade at Kohima, the bloodiest battle of the war in Burma. Like Cassels, Mike West always drove his own jeep, and he was never seen anywhere without a long shepherd's crook. In spite of his considerable bulk, he was as agile as a mountain goat across country. The job of ADC to Mike West was no sinecure, and one of his aides, who was some twenty years his junior, ruefully described a tour of the forward positions to an unsympathetic audience: 'He said he was going for a stroll, and we walked twelve miles.'

Korea was no place to command a division by telephone, and Mike West had a deep-rooted hatred for the instrument, which he still has. Consequently, no one in the Division knew where the tall, long striding figure with the shepherd's crook was going to appear next.

In the summer and autumn of 1952 Korea had become essentially a 'subalterns' war': there were few major operations at battalion or even company level, but hardly a day or night passed without a platoon clash, an ambush, a brisk and bloody fight in no-man's-land. General West made the cynical observation: 'If everybody below the rank of major would go home, this show would have a chance of going off without a hitch.'

As always when a war becomes bogged down in static stalemate, there were clamorous demands from the Higher Command for prisoners. By 1944, in Burma, the number of Japanese captured unwounded numbered about a dozen; in Korea in 1952 the figure was scarcely more encouraging.

Periodically, however, a prisoner *was* taken – usually as a result of daring 'snatch' raids at night. Competition to bring in a prisoner became fierce, and rewards of bottles of whisky and an extra five days R and R leave were an added incentive. High scorers were the Canadians; less impressive was the record of the Australians, who on several occasions tended to forget the object of the exercise and charged with fixed bayonets. The Welch and the Leicesters achieved one each, and a fierce argument developed between 'B' and 'C' Companies of the KSLI: eventually the prize was awarded to 'B', because it was decided that 'C' had done nothing more than apprehend a deserter. To the chagrin of the 1st Norfolks, their three prisoners died on them: they were all only slightly wounded, but like dogs curling up for sleep they expired for no attributable reason, and two privates found themselves on a charge as a result – being done out of R and R leave was bad enough, but to be put on the peg because a couple of Chinks died was 'tearing it a bit' . . . The

Black Watch took one, but he tried to escape and was shot dead. It was all very unfair . . .

Meanwhile, the 'peace' talks continued, but to no avail. Peace was not appreciably nearer . . .

[5]

In the autumn of 1952 the 'battles for the hills' started up again. Although there was no noticeable increase in activity on the front of the Commonwealth Division, the Chinese launched a number of furious assaults against the 1st Marine Division on 'Bunker Hill'; the 3rd United States Infantry Division on 'Big Nori' and 'Kelly Hill'; the 7th United States Infantry Division on Triangle Hill; the 25th Infantry Division on 'Heartbreak Ridge'. For the American GI the Korean War was very definitely still on . . .

[6]

In Korea a soldier could get to like a hill or hate it worse than hell: the 1st Middlesex, for instance, had a certain affection for Plum Pudding Hill – they had chased the North Koreans off it in double-quick time, and had managed to make themselves reasonably comfortable there for a short period; 'Spike' Pike's pioneers had a feeling for Gloster Hill, which had been almost akin to love before the rest of the Battalion and the Chinese moved on to it; the KOSB had stormed Hill 355 and held it against 6,000 enemy – after a trial of strength such as this you could get to like a hill; the Argylls had not disliked Hill 282, until they had been scorched off it by napalm. The 1st King's rated the view from Yong Dong ('the Dong') as the finest in the world.

In the American sector there was a hill of a certain scenic shapeliness, which was known as 'Betty Grable'; there was another, oddly shaped like a female bosom, which the Marines called 'Jane Russell'. No one liked 'The Brown Bastard'.

But there was another hill, so dominating in its position, that both sides wanted it with a passion amounting almost to mania:

it was shaped like a horse shoe – huge, towering, menacing, raw-boned, indescribably sinister. First World War men speak of Hill 60 with a certain awed reverence, as do another generation of soldiers of Cassino. Korea veterans of the Black Watch, King's, and Duke of Wellington's Regiment have their own hill to bate the breath.

This is the hill known as 'The Hook'.

The 7th United States Marines knew all about The Hook: they had stormed it in the face of murderous fire and they had captured it with their customary *panache*. Having taken it, they were going to hold it, because the Marines never give ground. To the Marines, The Hook was Iwojima and Okinawa all over again: they did not actually hoist the Stars and Stripes on The Hook because there was no time, but Lieutenant-Colonel Mike Delaney, Commanding Officer of the 7th Marines, had given his instructions to the 'Wire Chief' – the signals sergeant, in British Army nomenclature – and the message was there for the United Nations Command to read: 'The 7th Marines have The Hook.' There was a sigh of relief all round: if the Marines had The Hook, then the Black Watch, the King's, and the Duke of Wellington's Regiment could start heading for home.

The Marines had The Hook, but they were not to retain it for long: they were subjected to the same treatment as that meted out to the KOSB on Hill 355, and they lost it: they were blasted out of their shallow trenches; they fought for every yard of ground with bayonets and rifle butts; near suicidal three-man counter-attack squads went in with flame-throwers: no men of any nationality have ever fought better before or since, but the sad fact remained that by the afternoon of October 27th, 1952, The Hook was once more in Chinese hands. The 7th Marines had lost a position, but they were going to lose no time in telling the world at large that they had got it back again.

And they did: during the afternoon of October 27th, supported by one of the biggest artillery concentrations of the entire Korean War and with every available aircraft swooping down on the Chinese on The Hook, the 7th Marines stormed back into the attack. They lost close on two hundred men doing it, but The Hook was once more United States Marines' property. And this time they were going to keep it: Colonel Delaney made that very clear to his company commanders the following morning.

'The Hook,' declared Delaney, 'is the property of the 7th Marines, and just let any sonofabitch say different.'

No one had the temerity to deny this fact, but it was soon clear that the 7th Marines had had more than enough of The Hook.

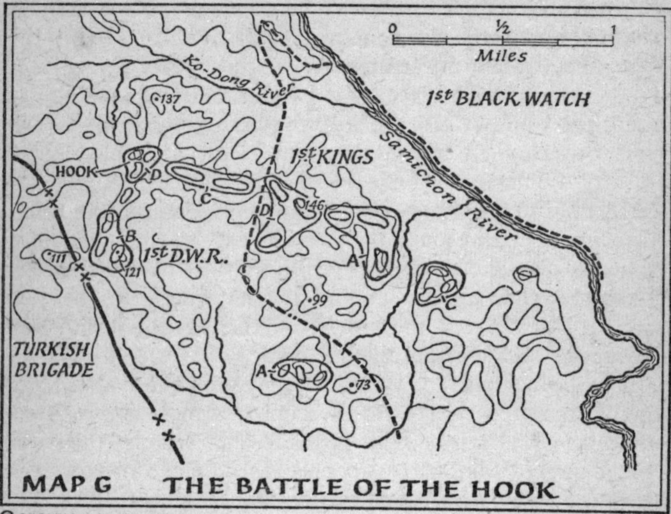

© CASSELL & CO LTD 1967

And this fact was made very apparent to Lieutenant-Colonel David Rose, Commanding Officer of the 1st Black Watch, when he effected a relief of the position on November 14th.

For Lieutenant-Colonel David Rose, Commanding Officer of the 1st Black Watch, the takeover of The Hook was scarcely a reassuring occasion, although the 7th Marines strove to convince him that all was well on the position. American infantry invariably left a position with the maximum possible speed, pausing only to negotiate a swift trade for liquor, but the Marines always departed with a more proprietorial air – the Marines took and held this position, they seemed to say, and it's up to you guys to see that it stays taken and held. It therefore came as something of a rude shock to David Rose to hear that the Marines gave them just twenty-four hours on The Hook – 'and that's stretching it a piece,' said Colonel Delaney with a pessimism foreign to a United States Marine; 'It's my guess those bastards will have it off of you by this time tomorrow. Meaning no disrespect to your boys, of course.' To David Rose this pronouncement came as a nasty surprise: Brigadier Joe Kendrew had told him, 'I'm giving you The Hook, David, and I know you'll hold it.' In receipt of such an order, there was nothing more to be said – at any rate not by the 1st Black Watch.

It was immediately clear to David Rose when he first saw The Hook, that the 7th Marines had had just about as much as they could take: the clean-shaven and sunburnt faces of the Jocks contrasted oddly with the drawn, exhausted and bearded American ones.

In the Command Post, Colonel Rose learned a little more about The Hook and it made grim hearing. 'We're real glad to see you guys,' Delaney told him; 'this is just about the goddamndest position I ever struck, and I did three years in the Pacific. But the brass say it's gotta be held, so I guess it's just gotta be held. So far we've managed to keep the gooks out. But my boys, what's left of 'em, are bushed – they haven't averaged two hours' sleep a night in three weeks.'

Somewhere to the rear artillery opened up harassing fire; away to the right a machine gun stuttered spitefully. 'It's been rough, Dave,' said Delaney – Colonel Rose winced slightly for Christian names were not usually introduced this early in the British Army – 'Look, I sweated out Guadalcanal and Saipan, but I never came a half so near as getting my ass shot off than I have in this goddam dump. That's how rough it is. You see some combat in the war, Dave?'

'Here and there,' said Rose, who had the Distinguished Service Order and four campaign stars.

'Well, you'll soon see I'm not kidding. During the daytime

there's shelling and mortaring and you can't do a thing except sit an' take it. They put their infantry boys in at night and, hell, those guys don't *want* to live. Well, I do, and it looks like I still might.'

David Rose, like every other British officer in Korea, was a staunch admirer of the United States Marines, even though they sometimes had deprecatory things to say about American infantry. Rose's experienced tactical eye told him at once that there were grave defects in the Marines' system of defence; like the American infantry, the Marines only seemed to have studied the technique of attack.

Clearly this was neither the time nor the place to tell a United States Marine Colonel his business, but to Rose it seemed that the defensive line consisted virtually of one long and shallow trench, known as the main line of resistance, which snaked its way up hill and down dale; forward of this tenuous ditch – there was really no other word for it – were a few outposts; to the rear could be seen a profusion of tents. Colonel Rose had first learned about 'defence in depth' twenty years earlier at Sandhurst, and had been both preaching and practising it ever since.

David Rose viewed the position once more, as he formed his concept for fighting a battle on it. He had heard how the Marines had lost The Hook, and how they had only managed to re-take it with heavy casualties, albeit displaying enormous gallantry. Rose resolved there and then to avoid having to undertake a counter-attack manoeuvre: it might be expensive in human life to hold the position at all costs, but infinitely more so to lose it and regain it by counter-attack: David Rose wanted no blood bath; no repetition of Vimy Ridge – if he was going to win a bar to his DSO (as he subsequently did), it was not going to be at the cost of a hundred dead Jocks. It therefore came as no great surprise to company commanders of the 1st Black Watch to learn that the position 'would be held to the last man and the last round'.

To one United States Marine, at least, the relieving Black Watch came as a grave disappointment. A gigantic Negro turned to his buddy, a white marine, in the foxhole, and said in tones of hurt surprise: 'Someone's done fooled us about these guys of the Black Watch.'

'How's that again, Sam?'

'Thought I mighta made a transfer,' said the Negro marine; 'the way I heard it, only the officers was white – the enlisted guys was all coloured boys like me. *Black* Watch – get it?'

'Ya mean some kind of a Negro outfit?'

'That's what I meant. But man, there ain't a coloured guy to be seen.'

'That suits me just fine,' said his friend.

The Jocks viewed the inhospitable terrain with morose fatalism: even the most obtuse private knew that hours of back-breaking digging lay ahead: taking over from the King's Liverpool had been another story – they had taken over a position which resembled a below street level bar in Sauchiehall Street.

'What like's this one?' demanded the Jocks of the outgoing marines.

'She's a real sonofabitch,' replied the marines, and they spoke from the heart. Soon the Jocks saw what they meant . . .

With Marine Colonel Delaney's gloomy prophecy to guide him, Colonel Rose launched the 1st Black Watch into the toughest programme of digging and wiring yet experienced by any British battalion in Korea. The defences had been terribly battered and were completely uninhabitable when the Marines left the position.

It was a question of 'dig or die': the Marines, in common with other American fighting men, had not dug and many of them had died. The relatively low rate of Black Watch casualties on The Hook was almost entirely due to the hard work of the Jocks on repairing the defences.

Wiring parties erected an elaborate defensive pattern; communication trenches and fighting bays were constructed; Royal Engineers, who regarded the infantry with tolerant condescension, constructed circular tunnels into which the troops could retire under heavy shelling and from which they would emerge to deal with the enemy when the shelling had lifted.

The work was only just completed in time. For on the night of November 18th, the Second Battle of The Hook started in real earnest . . .

[7]

The Hook was held, because the Black Watch had officers like Major Angus Irwin. The Jocks cursed Irwin from hell to breakfast time as their entrenching tools sank into the crumbly soil and company sergeant-majors swooped menacingly on men who were taking time off for a smoke. But they thanked him in the end: Irwin, too, was an admirer of the United States Marines as fighting men, but was filled with horror at the state of the position which he had been charged to hold to the last man and the last round.

The position presented a scene of indescribable devastation: the few bunkers that the Marines had constructed were smashed to matchwood; communication trenches, which should have been at least six feet deep, were scarcely knee high. Shredded sandbags and smashed pieces of barbed wire littered the area, and fragments of men, both American and Chinese, rounded off a scene which rivalled the illustrations in Dante's *Inferno*.

The tempo of the work and the rigid shellfire discipline imposed was such that when the first of many shells fell on 'A' Company not a single casualty was suffered. Jocks who had been complaining bitterly about the tempo of the digging programme were forced into the reluctant admission that Angus Irwin had probably saved their lives.

The Hook was essentially a 'subalterns' battle', and it was held because the Black Watch had subalterns like Second-Lieutenant Roger Doig, who died leading his platoon in a bayonet charge into a veritable hail of machine-gun fire; like Second-Lieutenant Michael Black, who although repeatedly attacked and surrounded for the whole of one night, sent back reports so cheerful and reassuring that Major Irwin was constrained to ask him on the wireless if he was really as happy as he sounded.

Lieutenant John Moncreiff, the Battalion Intelligence Officer, intercepted the following conversation between Irwin and Black, which was notable for its admirable economy in words:

> IRWIN: Are you happy where you are?
> BLACK: No, never have been.
> IRWIN: Don't be a clown. Out.

This was the nearest to a complaint that Michael Black, in a situation which was sticky always and frequently near desperate, ever ventured except once towards dawn on November 19th:

> BLACK: Some enemy are throwing grenades down on us.
> IRWIN: Without being facetious, throw some back.

The Hook was held because there were sergeants like Alexander Hutchison, who, when forced to his knees by a severe wound, led his men into the counter-attack on his hands and knees; corporals like Robert Manning, who refused to be evacuated after being twice wounded and who silenced an enemy machine gun with a hand grenade.

Privates like George Coley, who bowled Chinese over like ninepins with his Browning machine gun up to the very moment when he was killed by a grenade; like MacDonald (16) and Alex Graham, who became prisoners-of-war, but only for very brief periods: MacDonald, deprived of his rifle by his captors, engaged them with his fists and made good his escape, while Graham resorted to even more lethal methods – he split the skull of the Chinaman guarding him with the brim of his steel helmet.

A battalion's war diary, kept in the heat of battle, is rarely a literary masterpiece: Adjutants and Intelligence Officers, with shells falling all round them and telephones buzzing incessantly, rarely commit more words to paper than are strictly necessary.

One entry in the battalion log of events, however, expresses the spirit of the 1st Black Watch in the defence of The Hook. This entry reads as follows: 'Serial 130. Time 0558 hrs. CO requested a Sitrep from "A" Coy. Sitrep: Situation very confused. Small parties of the enemy have suddenly appeared all over the shop. CO asked: "Have you any reserves left?" OC "A" Coy replied: "No." CO said: "Stick it out."'

And the 1st Black Watch did . . .

SIXTEEN

At long last, after a lengthy recess since October 8th, 1952, it looked as if some good might emerge from the peace talks. The atmosphere of the talks was scarcely more cordial, but it was noticeable that the Chinese delegates were less deliberately obstructive and even occasionally agreed with a statement made by an American: this may have been due to the fact that the Chinese knew that they could not win the war – their recent offensives, particularly that made on the Black Watch on The Hook, had been conspicuously unsuccessful – or that the American delegates were learning tougher tactics at the conference table.

Whatever it was, the exchange of sick and wounded prisoners, known as Operation Little Switch, began: in the week between April 20th and 27th, the United Nations handed over 5,194 North Koreans, 10,030 Chinese, and 446 North Korean civilian internees. The Communists, as if anxious to prove that they were willing to cooperate, replied with 471 South Koreans, 149 Americans, 32 British, 15 Turks, 6 Colombians, 5 Australians, 2 Canadians, 1 Greek, 1 South African, 1 Filipino, and 1 Belgian. It made an unimpressive total, but at least it represented positive action, which so far had been conspicuous by its absence.

[2]

In a United Nations prisoner-of-war compound in North Korea, men of many nationalities heard the news of Little Switch with wry disbelief. Many of them were wounded and all of them, to a greater or lesser degree, sick: those captured in the early fighting had endured two dreadful winters in unspeakable conditions, on a diet of rice, anaemic cabbage, potato soup and dried beans.

In this particular compound there seemed to be a number of likely candidates for exchange: there was an American infantry lieutenant who had lost a leg, and another whose heart had become weak; there was a major, crippled with beri-beri, who would assuredly die if he stayed in the camp for another month; there was a Turkish sergeant with a terrible, gaping wound in his right thigh and buttock, which obstinately refused to heal and was rapidly becoming gangrenous; there was a South African pilot, whose wounds necessitated immediate and extensive surgery (he was, in fact, the single South African referred to earlier).

Obviously very low on the priority list were three officers of the Gloucesters – Major Denis Harding, Captain Anthony Farrar-Hockley, and Captain 'Spike' Pike: Harding had been tied to a beam by his wrists, forced to stand stripped to the waist in the snow for twelve hours and sentenced to six months' solitary confinement for refusing to sign a 'confession' of 'conspiracy'; Farrar-Hockley, with three escape attempts behind him (he was busily formulating plans for a fourth when the news of Little Switch came through) was considered, reasonably enough, by the Chinese to be the *enfant terrible* of this or any other camp; 'Spike' Pike, whose attitude to Communism had been consistently 'incorrect', had been beaten up and sentenced to solitary confinement for inciting a labour strike.

Captain James Majury of the Royal Ulster Rifles was not over hopeful, even though he had contracted tuberculosis: in the absence of a priest in the compound, he had taken over the job of lay Protestant Chaplain, while Captain Jumbo Wilson of the Gloucesters performed the same office for the Catholics, and the worship of God had never been popular with the Chinese.

The Chinese obstructed the Chaplains in the prison camps in every possible way: there would be no place, they said, for churches or God in the Great New Communist World.

The efforts of priests to perform their ministry were greeted with disgusting cynicism and puerile attempts to mock their faith. Various prison-camp commandants confiscated Bibles and hymn books – with paper for domestic purposes at such a premium, it surprised them that the Chaplains did not turn Bibles and hymn books to better use.

There were a number of other men of 29 Brigade, captured in

the Imjin Battle, who were clearly 'low in the betting' for exchange: Fusilier Derek Kinne would normally have been selected by any doctor on sight – he had been kicked so savagely that he had sustained a double rupture and was unable to walk; a number of other equally unquenchable spirits had been manacled and confined in boxes, six by three by three feet, for periods of up to six months; there were several Ulster riflemen, and one of them was almost certainly the man who had told the Chinese lecturer to f— himself after the January 1951 massacre.

None of these men fancied their chances over much, and the story goes that a soldier who had lost a leg was selected, but ruined his chances of early repatriation by issuing the same bizarre request as the anonymous Ulster rifleman to a Chinaman when asked to sign a statement that he had been handsomely treated during his captivity.

And so the more fractious prisoners – sick, emaciated but unbroken – waited for the war to end . . .

[3]

With the improved atmosphere of the peace talks, it seemed that the armistice could not be delayed much longer.

But The Hook still remained a notorious hot spot. Like a frustrated speculative builder, constantly baulked of planning permission on a choice piece of land, the Chinese launched themselves at the hill: they gave all the sun's hours to the guns and mortars, and at night the infantry came, as always, head-on.

Americans who fought on The Hook grew to speak of it in awed tones: it became known as the foulest piece of ground in the world; it was rotten to the core; it stank of death; it was not possible to dig without encountering human remains; it was barely fit for the rats which foraged diligently in every trench. The American infantrymen spoke with feeling: between January and April 1953 The Hook changed hands something like twenty times.

The 1st Black Watch, invigorated by a celebration of Hogmanay (which had astonished even the Australians) and a period of rest and comparatively uneventful reserve duty, came back to The Hook on April 7th. Colonel David Rose had listened to one grim assessment of The Hook from Colonel Delaney of the 7th Marines, but the infantry colonel vacating the position – a picturesque character who spat at frequent intervals – painted an even blacker picture: The Hook was Guadalcanal, Iwojima, and Okinawa all rolled into one, and worse than any of them. Holding The Hook, it appeared, was still a rugged assignment.

The Black Watch had seen the enemy off The Hook in November and were fully confident of doing so again.

Just after midnight on May 8th, following a heavy daytime bombardment, the attack came in and the Jocks were ready for it: illuminated by flares and searchlights, the enemy were caught by the machine guns of No 3 Platoon of 'A' Company; bathed in the light of the flares and searchlights, they presented a target to set any machine gunner's mouth watering – the principal architects of destruction in what came to bc known as the Third Battle of The Hook were Privates Cash, Clark and Petrie who fired 8,000 rounds between them and claimed to have killed one hundred Chinese. 'I should hope so, too,' said RSM Bill Scott severely, 'for I doubt there's that amount of ammunition in the whole battalion . . .'

As in the November battle, the Jocks showed themselves fiercely averse to being taken prisoner. Private Roy Quinn, temporarily cut off from the rest of his section was 'jumped' by a Chinaman who leapt on his back. Quinn hit the Chinaman in the face with the butt of his rifle, and completed the job by beating his adversary's head against his knee. Another Chinese, coming to his comrade's aid, was disposed of by Quinn by a process which he modestly described as 'kicking him up'. Lieutenant John Moncrieff, the Battalion Intelligence Officer whose duty it was to collate such reports, kept a tolerably straight face during this recital. But there was more to come: three more Chinamen were approaching Quinn, and he decided that discretion might be the better part of valour – 'I didnae mind two Chinkies, but I thocht three was ower much.'

Private MacKenzie of 'B' Company also seemed to be in danger of being made captive, but mindful of a bottle of whisky and five days 'R and R' in Tokyo as a reward, decided to take one himself.

'I tried could I find a prisoner to bring in,' he reported to Lieutenant Moncrieff. 'A Chinky come at me – a big gowk – and I grabbed a hold of him and swingit him round. He brocht his heid doon as if for to butt me, so I made a grab for his hair, but it was cut too short and I couldnae hold on. So I gied him a skite in the lug wi' his ain gun.'

'But no prisoner?' queried Moncrieff, whose self-control was fast deserting him.

'He wouldnae be ta'en,' said MacKenzie with beguiling simplicity, 'so I kilt him . . .'

The Chinese had been denied The Hook by the Black Watch once more; their artillery had fired 575 shells on the Jocks, but had failed to shift them. Although certain official histories dismiss this battle as only a 'probe', the Black Watch call it the Third Battle of The Hook and fairly enough claim it as a battle honour. They were unable, however, to claim any live prisoners to the chagrin of half a dozen Jocks who produced excellent reasons for claiming five days 'R and R' or a bottle of whisky in lieu: two wounded Chinese were brought in, but emulating the prisoners of the 1st Norfolks, they died when the Company Sergeant-Major's back was turned for a moment. Colonel David Rose, whose own 'R and R' had been rudely interrupted, ruled that they did not count for leave, but qualified for a bottle of whisky.

At odd moments during the Black Watch tenancy of The Hook, a Chinese propaganda loudspeaker had been busy: 'Scotsmen, go home!' had been a favourite request, and a trilling feminine voice had actually given a rendering of 'Loch Lomond'; the Jocks had joined in with smoking concert fervour. But on the evening of May 8th, as the Chinese collected their wounded and slunk away from The Hook, the loudspeaker spoke again with spiteful emphasis: 'You only beat off a patrol action last night, but you will have something bigger to deal with in five nights' time.'

This was reported laconically by Lieutenant John Moncrieff in the Battalion's battle log, timed 2114 hours. The next entry was timed 2130 hours, but Moncrieff did not consider it worth a lengthy entry: 'Same again as above,' he wrote. 'The general theme was that they accepted that they were beaten up last night, but wait for five nights from now.'

John Moncrieff drew a thick black line under this entry, and prepared to vacate the Battalion Command Post: in five nights' time The Hook would be someone else's problem, and in the meantime he was long overdue for a bath . . .

[4]

The Black Watch passed the baby to the 1st Duke of Wellington's Regiment, and a singularly unwholesome infant it turned out to be.

'The Dukes' were by way of being a favourite of Brigadier Douglas (Joe) Kendrew, now commanding 29 Brigade. The reason for this was not hard to see: the Dukes have always been the premier rugger regiment of the British Army, and it is popularly supposed that pre-war and post-war gentleman cadets at Sandhurst, who showed outstanding talent on the rugby field, were fastened upon avidly by the Duke of Wellington's Regiment.

Joe Kendrew was a rugby star in his own right: on ten separate occasions he had been the mainstay of the English pack – he had been captain of the England Fifteen in 1935 – and in inter-service matches it was said that his shout of 'Feet, Army, FEET!' could be heard in the saloon bar of the Winning Post tavern. He had been equally vigorous as a soldier in war, and when he arrived in Korea already had three of his four DSOs – as battalion and brigade commander he had won the first in North Africa, the second at Salerno and the third at the crossing of the Meccia River. In Korea he was shortly to add a fourth.

The Hook, however, had an evil reputation, and at first it seemed, in rugger parlance, that to put the newly arrived Dukes on the position would be rather like pitting Rosslyn Park Extra 'A' against the All Blacks: the Dukes, like the Middlesex, KSLI, KOSB, and Leicesters, were predominantly National Servicemen; unlike these four battalions, whose name was secure in Korean War history for all time, they had not had the benefit of months of hard training in Hong Kong before coming to

Korea – they had come direct from leisurely garrison duty in Germany which, as any experienced soldier will tell you, is the worst possible training for active service.

But the Dukes did have the advantage of an extended 'shake-down' period in a quieter part of the line, and after two months were patrolling and doing without sleep like veterans. Early in their service in Korea, the Dukes experienced mortaring and shelling, and they very soon learned the truth of the dictum 'dig if you want to stay alive'. Before 1953 was one month old two young subalterns, Second-Lieutenants Rodney Harms and Patrick Orr, had won Military Crosses for daring raids on enemy positions in front of The Hook.

Spiritually, too, the Dukes had undergone a considerable change since the casual and unreflective military way of life which prevailed in Germany. In Germany the Reverend William Burns, Chaplain of the 1st Duke of Wellington's, had had his problems – some six hundred in all, and they had not shown themselves over receptive to spiritual medicine.

In Germany William Burns found the job of battalion Chaplain less rewarding than he had hoped: his weekdays were spent listening to involved stories of domestic betrayal, and on the Sabbath his congregations were depressingly small.

But in Korea it seemed that war had brought with it a spiritual awakening. The language of the bunker was the same as that of the soldier the world over; the usual untuneful and copulative adjective made its appearance in every sentence; the subjects under discussion were limited to alcoholic and sexual prowess. But it seemed to Williams Burns that soldiers who of necessity lived in close proximity to one another came closer to God at the same time.

That a full-scale attack was imminent, everyone knew. For three days, ever since May 13th – the date forecast by the Chinese propagandist on the loudspeaker for the final eviction of British soldiers from The Hook – the Dukes watched and waited: hands gripping weapons; nerves stretched to breaking point; eyelids drooping with the effort to stay awake.

It therefore came as something of an anti-climax when a small and unarmed Chinese, hands held high in the air and wearing a conciliatory smile, walked into No 8 Platoon's position on the morning of May 18th.

He was spotted simultaneously by three privates of the Dukes, who all claimed him as their prisoner – the reward of five days' 'R and R' leave was still in operation for the apprehension of a live and breathing Chinese soldier, and the Dukes had listened without sympathy to the sad tales of the Black Watch and Norfolks whose prisoners had died on them. But this one wasn't dead – very far from it: the three Dukes made a concerted grab for him, and the one who first laid hands on him shouted to his platoon commander in triumph: 'Got him, sir. How about that five days' Tokyo leave?' The platoon commander had been done out of 'R and R' leave himself, and the subject still rankled. 'Get him up to Company Headquarters,' he ordered, and cursed himself for not having had the presence of mind to snatch the prisoner first.

'How about some Tokyo leave, Sar-Major?' said the private, on arrival at 'C' Company Headquarters; one of his mates in the platoon had sampled the joys of 'R and R' and it sounded just what the doctor ordered, although the doctor, to whom he reported on his return, had thought otherwise . . .

'I'll give you "R and R",' said CSM Jack Randall grimly – he hadn't had any Tokyo leave either – 'bring him to the Company Commander, and look sharp about it.'

The Company Commander, Major Barry Kavanagh, was highly delighted at having a prisoner – the demands from Brigade Headquarters had been very insistent during the past week – but he hardly expected one so quiescent and immediately cooperative.

Previous prisoners captured by the Dukes – they numbered only two to date – had been hangdog, surly and inarticulate, and irresponsive to the somewhat primitive interrogation methods of the South Korean interpreters attached to the battalion, which usually consisted of a slap here and a well-aimed kick there as an aid to the memory. The first two prisoners had said nothing of any use, or more likely, had known nothing. But this one might have emerged straight from the Chinese divisional commanders councils: indeed, he might even have been the divisional commander himself, so exact and painstaking was his memory.

His name was Hua Hong, and his rank was private: the Chinese may have wanted The Hook with a fanaticism bordering on the insane, but Hua Hong had a very much less exalted ambition – he wanted a cigarette and an immediate discharge from the Chinese Communist Forces, and he got both.

Hua Hong was not untypical of the Chinese soldiers operating in Korea at this time: he did not want to be a soldier, he felt no enmity against anyone (except the Chinese High Command who had conscripted him into the army) and the principal worry was the fate of his family at home – in these respects he was not unlike the less militant National Servicemen of the British Army.

For a Chinese peasant, his intelligence was about average, although he was completely illiterate. He hated the Chinese Communist Forces at least as much as the most unwilling conscript hated the British Army, and his daily routine was not appreciably different: from 6 PM in the evening until 4 AM the following morning he dug and constructed fortifications and went on patrol; he had his breakfast, consisting of rice and millet – the first of two daily meals – at 4 AM, and slept from 5 AM until midday. Apart from the rice and millet, never favourite items on a Duke's menu, the programme seemed conventional enough.

But Hua Hong's free time – something like five hours in every week – was not his own to write letters home, read ancient magazines, play pontoon and re-live nights out in Minden. These five hours were spent in political instruction and indoctrination, and inattention or lack of enthusiasm in this direction was immediately reported to the unit political officer with dire results (a British soldier could refer to the CSM as a fascist bastard to his mates without fear of repercussions; in a Chinese unit every word got back).

Like the average CCF soldier, Hua Hong was a combination of paradoxes: he was afraid of the more extreme Communist members of his unit, but fought bravely alongside them; his training, by British or American Army standards was poor, but he was remarkably skilful when fighting in a small group.

Thus, Private Hua Hong of the Chinese Communist Forces was soon, under the influence of a Lucky Strike and a mug of company office tea, giving Major Kavanagh information which a small army of specialized intelligence officers had been trying to obtain for weeks.

The information given by Private Hua Hong was that an all-out attack on the Dukes on The Hook was imminent, and this time the Chinese were coming to stay; the assault would be made by five companies of infantry, and a further three companies would take over the captured positions and hold them; if Hua Hong's figures were reasonably accurate, the Dukes would be outnumbered by something like five to one.

Hua Hong seemed to know everything about the forthcoming attack on The Hook – except the date . . .

Colonel Francis Bunbury was not the most gullible of men, as many Dukes ready with convincing stories to explain absence without leave had discovered to their cost. But he believed Private Hua Hong's story, and immediately resolved to make The Hook as nearly impregnable as possible: during the next fortnight he set about deploying his battalion to the best possible advantage and strengthening the defences with all possible speed. The 1st Duke of Wellington's Regiment, declared Bunbury in a forceful directive to his company commanders, were not going to be caught napping – it goes without saying that he threw in a governing clause about the last man and the last round, because in Korea this had become standard procedure.

Colonel Bunbury may have had certain qualms about the fighting capabilities of his battalion on arrival at Pusan, but by May they had completely disappeared. After the Battle of the Alma, the Colonel of the 33rd Regiment of Foot (later the Duke of Wellington's Regiment) told his men on September 20th, 1854: 'You are a credit to your country and the Army; for I would take you into battle and be sure of winning it.' Nearly one hundred years later Francis Bunbury was to address the same battalion in almost identical words . . .

Tension on The Hook mounted almost unbearably, and every night for the next fortnight the message 'tonight's the night' was passed from platoon to platoon. It seemed that the night of May

20th was indeed the night – from late evening until the dawn of the 21st 4,500 shells fell on the Dukes, and thereafter the enemy artillery averaged 2,000 shells every twenty-four hours on The Hook.

It would have been a terrifying ordeal for even the most battle-hardened troops, and this the Dukes certainly were not; like every other British battalion in Korea, they had a nucleus of officers and NCOs who were experienced in war, but every one of these veterans was adamant in his opinion that nothing that they had experienced in the Second World War rivalled the bombardment of The Hook. No officer or man in the Brigade had more fighting experience than Brigadier Joe Kendrew, and he described the shelling of The Hook as the worst in his experience.

'How are they sticking it?' asked Colonel Bunbury from his command post, and every company commander was able to reassure his colonel that their men were taking the hammering like veterans.

Francis Bunbury was worried, although not a single Duke would have guessed it as his tall and debonair figure appeared in every company position during the bombardment; he was worried because the shelling was becoming more than human flesh could stand. He knew, however, that the only thing to do was sit tight and take it – the infantry attack must come in soon.

But during the next three days the shelling and mortaring merely increased, and during this seventy-two hours an absolute fury fell on The Hook: there was a steady drain of killed and wounded; bunkers collapsed and men were buried alive; platoon command posts were receiving direct hits – a screaming mob of Chinese surrounding the position would have come as a positive relief, if only the shelling would stop. But the Dukes had been told that they would stay on The Hook; if they could not stay on it alive, then they would stay on it dead.

At 8 PM on the evening of May 28th the bombardment lifted just perceptibly, and the Chinese were at the Dukes' throats: from that moment there was not an officer or man of the battalion, save those dead, wounded or buried alive, who was not fighting desperately for the remainder of that fearful night.

The fourth (and last) battle of The Hook resolved itself into a series of desperate close-quarter engagements fought at a range varying from twenty yards to a few feet. One of the most violent and bloody of these was fought by Major Lewis Kershaw and ten men, who can be said to have explored the nethermost regions of hell.

Kershaw was in one of the forward platoon posts, which received three direct hits from heavy shells – there were men killed outright by blast; men blown into small pieces; men stunned, deafened and blinded. Anticipating the collapse of the command post by a mere sixty seconds, Kershaw ordered those men still on their feet – some ten in all – to take shelter in a nearby tunnel: he was the last to enter the tunnel, and no sooner had he done so than a party of Chinese appeared at the entrance and threw grenades inside; before falling seriously wounded, Major Kershaw dispatched two of the Chinese with his sten gun. For Lewis Kershaw this was the beginning of months of agonizing pain, culminating in the amputation of one of his legs, but he fired at the Chinese until his sten magazine was exhausted. He was not yet done with the Chinese: although unable to stand, he threw grenades from a lying position and prevented the entry of the enemy into the tunnel. As dawn broke on the 29th he was still there, together with ten wounded men, and the tunnel was still Dukes' property . . .

There were many other Dukes of similar calibre: Corporal George Pickersgill denied the Chinese entrance to another tunnel – practically blinded by the blood which poured down his face, he poured bursts of fire into enemy crowding the entrance; Private Denis Husband, blocking yet another tunnel, was invited by the Chinese to surrender – he replied with an oath so frightful that afterwards he could not repeat it even to himself, and a shower of grenades. He was subsequently buried alive by mortar fire, and on being dug out next morning demanded to be let loose against 'them f— chinks' again.

The Hook was held because the Duke of Wellington's Regiment had such soldiers. On the morning of May 29th not a single bunker was undamaged on The Hook, and practically every trench had been levelled. On the position were 24 dead Dukes, 105 wounded Dukes and a number still buried under the rubble.

Of live Chinese there were none, but there were 250 dead – some spreadeagled on the wire, some cut down by fire only a few feet from the forward positions, some in fragments inside the battalion perimeter, blown to bits by defensive artillery and mortar fire. It would be a long time before the air was clean on The Hook.

Little squads of grimy Dukes stared blankly at the fresh and perky young cockneys of the 1st Royal Fusiliers as they took over The Hook: many were too dazed to realize that they had been relieved, and too tired to believe their sleep-starved eyes.

'Nice place you've got 'ere,' said the Fusiliers.

'It's all yours, mate,' said the Dukes grimly.

'The Dukes,' said Brigadier Joe Kendrew, 'held The Hook, as I knew they would.'

SEVENTEEN

It looked as if July 27th, 1953 was going to be like any other summer day in Korea: it was stinking hot, and the sun beat down fiercely on the brown scrub and dusty, ochre-coloured soil on the hills of Korea. The clear sky shimmered with the heat, and the fine brown dust was everywhere – on the telephone wires, on parked vehicles, in the working parts of weapons, in the hair, down the throat.

There was little activity on the British Commonwealth Division front, but the constant grumble of mortar and shellfire and the nagging chatter of machine guns were grim reminders that the war was still on, and that unlucky and unwary men could still die.

The 1st Royal Fusiliers still strove to make The Hook habitable, and received their baptism of fire as they worked; the Durham Light Infantry were patrolling busily, and only a few nights ago had fought a neat little action which had resulted in a dozen dead Chinese; the 3rd Royal Vingt-Deuxième Regiment had killed twenty more, for the loss of one man slightly wounded; the 2nd Royal Australian Regiment, recently arrived in Korea, were getting restive – it seemed possible that the war might end before they got the chance of a crack at the Chinese.

The 1st Duke of Wellingtons, licking their wounds – they suffered 126 killed and wounded on The Hook – reclined at ease on the summit of a lush hill overlooking the Imjin River, and regaled the incoming 1st North Staffords with terrible and partially true tales of battle; the 1st Black Watch had handed over to the 1st Royal Scots, and were at sea on their way to Kenya, whence dark deeds were reported.

There were rumours that the peace talks were going well, and that an armistice might soon be signed; the soldiers had heard this many times before – the only occasions for rejoicing were if you saw your name on the company detail as being due for home or 'R and R' leave in Tokyo. In the meantime, life was tolerable if you weren't being shelled and it wasn't raining. But most of the time one or both these things were happening.

For the 1st United States Marines Division, in the line to the left of the Commonwealth Division, it was very much business as usual: since July 24th they had been subjected to ceaseless shelling and furious infantry attacks in fighting as savage as any in the whole war. That was their bad luck, but the men of the Commonwealth knew that it could just as easily happen to them, and the 2nd Royal Australian Regiment rather hoped that it would.

In the 'Peace Village' of Panmunjom, too, it seemed that July 27th was just another day – a long and frustrating day of stalling, bickering, accusation, and denial: the same bored looking American military policemen in their 'snowdrop' white helmets; the same scruffy North Korean sentries with their poker faces and long bayonets fixed; the same gaggle of pressmen, pecking avidly at any titbit of news which emerged from the drab tent which housed the American and Chinese delegates.

But July 27th was not to be just another day of futile and acrimonious negotiation: at 10 AM, in a stony and unsmiling silence, the Armistice agreement was signed – three years and one month after hostilities began, two years and eleven months from the time the first Commonwealth troops had landed at

Pusan, and two years all but a day after the formation of the First British Commonwealth Division.

At 10 PM that night men of the Commonwealth Division assembled on their positions in full view of the enemy, and exactly on the hour regimental buglers sounded the Regimental and Company calls followed by the 'Cease Fire'. A roar of cheering started on the right of the line: it was taken up by other units, and it gradually spread along the entire length of the line in a roaring crescendo of sound.

Red, white, and green Verey lights shot up into the night sky from every platoon position, producing a firework display that any schoolboy would have given his right arm to have seen. Officers and men cheered, whistled, and shouted; carefully hoarded liquor supplies were opened, and majors, subalterns, sergeants and privates toasted one another in tin mugs full to the brim with neat spirits.

The Chinese response was disappointing that night, but they made up for it the next morning when they decorated their positions with flags, paper streamers and lanterns. Propaganda broadcasts invited exchange visits across no-man's-land, but the time for fraternization was not quite yet . . .

In 'Freedom Village' the stage was set for the commencement of Operation 'Big Switch' – the exchange of prisoners-of-war. In a large building, hung with the Stars and Stripes, the Union Jack, and the blue and gold flag of the Commonwealth Division, soldiers bustled to and fro, ensuring there should be no hitch, for on the personal orders of General Mike West, returning Commonwealth prisoners were to get VIP treatment.

There was a gigantic tea urn, which represented the first stop for every man; there was the delousing department, where American medics were filling contraptions like fire extinguishers with white powder; there were doctors in snowy white coats, standing amid a strong smell of disinfectant; there was a clothing store, where the hated Chinese prison garb would be shed and subsequently burned; there were shapely Red Cross girls placing menus, on which steak and beer figured prominently, on tables surmounted by white cloths.

General Mark Clark stood under the Stars and Stripes; by the Union Jack was General Mike West, complete with shepherd's crook. Outside, those indefatigable Negro musicians who had greeted every soldier at Pusan played *Oh, What a Beautiful Morning*, *Waltzing Matilda*, *The Star Spangled Banner*, *The Maple Leaf Forever* and, inevitably, *Colonel Bogey*. Flash bulbs exploded in all directions.

The first man into the reception room presented an extraordinary spectacle: he was a small man, not more than five feet four inches tall, and he was encased in shapeless Chinese clothes several sizes too large for him. But the rakish angle of his quilted cap was unmistakable, as was his gesture as he removed the cigarette from his mouth with thumb and forefinger and spat out a shred of tobacco. His arrival resembled the sudden and unheralded appearance of Norman Wisdom in the House of Commons at question time. His query to an American major – 'Which way to Blighty, mate?' – was clearly audible at the far end of the room.

Then he caught sight of General West and the giant tea urn simultaneously; he threw down his cigarette end, ground it into the floor with a stockinged foot from which a big toe poked defiantly, adjusted his deplorable head-dress and snapped to the salute. Mike West returned the salute gravely, but it was some time before he could trust himself to speak . . .

One of the later arrivals at Freedom Village was Lieutenant-Colonel Fred Carne of the Gloucesters: he had lost three stone in weight, his face was gaunt and drawn, his tired eyes were incredibly deep in their sockets. He was immediately surrounded by a milling throng of Gloucesters, all trying to shake his hand at the same time. Then he was hoisted shoulder high and chaired through the camp.

Fred Carne has never been promising material for press interviews, and on this occasion he ran true to form. Asked about his period in captivity, he replied: 'The food was rotten, and I was damned bored.'

He had been in solitary confinement for nineteen months.

There was no nation-wide rejoicing at the end of the Korean

War; no frenzied displays of tickertape in Broadway; no one fell into the fountain in Trafalgar Square. The reaction of the free world was like that of a man who has had a painful, abscess-ridden tooth pulled out of his head at last.

In three years and one month of war the whole of the North Korean countryside was laid waste, and South Korea was shattered. Something like 400,000 homes were destroyed, and approximately four million people were rendered homeless.

Although the total casualties were scarcely more than those suffered by both sides in the Battle of the Somme, an estimated total of 1,820,000 men were killed and wounded in Korea, of which 1,420,000 were Chinese and North Korean. The South Korean army lost 415,000 dead and 429,000 wounded. American battle casualties amounted to 33,629 dead and 103,284 wounded. Comparisons are inevitably odious, and it may seem that the British Commonwealth casualties were negligible by comparison (in 1917 a Canadian Brigade lost more men in two days than the whole Canadian contingent in the Korean War): 93 officers and 1,170 other ranks killed; 301 officers and 4,516 other ranks wounded; 60 officers and 1,128 other ranks missing – of this last category, 40 officers and 996 other ranks, nearly all from 29 Brigade and captured in the 1951 Imjin Battle, were repatriated.

Many men of other nations poured out their blood on Korea's melancholy soil: Belgium, Colombia, Ethiopia, France, Greece, Holland, the Philippines, Thailand, and Turkey. These countries between them suffered 1,800 killed and 7,000 wounded; of these, nearly half – a total of 3,130 – were Turks.

There can be no neat and tidy end to a story about the war in Korea, any more than there can be a neat and tidy end to Communism. The Korean War can be called a good war, if such an adjective can be used to describe a conflict in which close on two million men were killed, crippled, blinded, rendered insane, disembowelled, beaten or starved to death, blown into small pieces, cut in half by machine-gun fire, burned, scorched or decapitated. It was a good war, because, although there was no victory, the United Nations triumphed over Communism and dealt it a deadly blow.

It temporarily halted Communism in the Far East, and staved

off a Third World War. But in 1967, the commissars are still on the rampage in the Far East, and boys from Texas, Minnesota, South Carolina, Alabama, Maine, and California are dying fast in their efforts to stop them. In Vietnam they are doing what their elder brothers did in Korea.

In 1967 quite a lot of the world is still free. But not enough of it.

Appendix A

CHRONOLOGY OF THE KOREAN WAR

1950

June 25th – North Korean People's Army crosses 38th Parallel to invade South Korea.

June 25th – UN Security Council calls for cease-fire in Korea and withdrawal of North Korean troops.

June 27th – UN Security Council adopts US resolution taking note of North Korean refusal to heed June 25th resolution and calls upon members to assist South Korea.

June 28th – North Koreans capture Seoul, capital of South Korea.

July 3rd – Inchon falls to North Koreans.

July 7th – General of the Army Douglas MacArthur appointed Supreme Commander of United Nations Command in Korea.

July 20th – North Koreans capture Taejon. Major-General Dean cut off and later taken prisoner.

July 30th – Beginning of Pusan Perimeter defence.

August 1st – Russia ends boycott of UN Security Council as Delegate Jacob Malik assumes Council presidency.

August 4th – North Koreans establish bridgehead across Naktong River.

August 4th – Soviet delegate Malik calls Korean fighting an 'internal civil war' and demands withdrawal of 'all foreign troops from Korea'.

September 1st – North Koreans open all-out offensive against Pusan Perimeter.

September 15th – United Nations forces land at Inchon.

September 17th – Kimpo airfield recaptured by UN.

September 19th – Enemy forces at Pusan begin collapsing.

September 25th – North Koreans in full retreat and UN pursuit begins.

September 28th – Seoul recaptured by UN troops.

September 30th – Communist China's Foreign Minister Chou En-lai warns: 'The Chinese people will not supinely tolerate seeing their neighbours being savagely invaded by the imperialists.'

October 1st – ROK (South Korean) troops cross 38th Parallel in pursuit of retreating North Koreans as General MacArthur calls upon enemy to surrender.

October 2nd – Chou En-lai informs Indian ambassador in Peiping that if US troops enter North Korea, Communist China will intervene in Korean War.

October 7th – UN General Assembly authorizes UN forces to pursue the enemy across the 38th Parallel.

October 10th – Wonsan on the east coast captured by ROK troops.

October 10th – Communist Chinese Foreign Affairs spokesman repeats warning of intervention in Korean War.

October 16th – First Chinese Communist troops – the 'People's Volunteers' – secretly enter Korea from Manchuria.

October 19th – Pyongyang, capital of North Korea, captured by UN forces.

October 26th – Chinese Communist troops attack ROK units at Yalu River and points south of the Sino-Korean border.

November 1st – First Russian-built MIGs appear along Yalu to attack US airplanes.

November 2nd – Chinese Communists with North Korean allies strike at United Nations units on western and eastern fronts.

November 5th – General MacArthur notifies United Nations that Chinese Communists are operating in Korea.

November 7th – All Communist forces in Korea break off action.

November 24th – General MacArthur announces 'win-the-war' offensive.

November 26th–27th – Communist forces, now controlled by People's Republic of China, attack on both fronts; deep penetration around Tokchon threatens to turn 8th Army's right flank, in the east 1st Marine Division flanked and cut off at Chosin Reservoir.

December 4th – United Nations forces in full retreat, Pyongyang recaptured by the Communists.

December 15th – United Nations forces begin establishing defensive line approximating the 38th Parallel.

December 27th – Lieutenant-General Walton Walker, commander of 8th Army, killed in road accident; Lieutenant-General Matthew Ridgway named his successor.

December 29th – Communist troop build-up reported above UN line at 38th Parallel.

1951

January 1st – Communist forces launch all-out offensive against United Nations; Ridgway begins orderly retreat.

January 4th – Seoul again captured by Communists.
January 7th – UN forces set up new defence line at Pyongtaek-Wonju.
January 15th – Enemy offensive halted.
January 21st – General Ridgway issues orders for counter-offensive.
February 7th – Communists forced to withdraw north of the Han River.
March 14th – Seoul recaptured by UN forces for second time.
March 31st – Leading elements of United Nations Command reach 38th Parallel.
April 11th – President Truman relieves General MacArthur as Supreme Commander of UN forces and replaces him with General Ridgway; Lieutenant-General James Van Fleet appointed as commander of US 8th Army.
April 23rd – Chinese Communist forces launch another offensive.
April 30th – UN forces complete withdrawal to prepared positions north of Seoul.
May 3rd – Enemy offensive halted.
May 16th – Second phase of enemy offensive given impetus.
May 21st – United Nations Command launches counter-offensive which succeeds in driving enemy north of the 38th Parallel.
June 13th – UN forces capture Chorwon and Kumhwa in the Iron Triangle.
June 23rd – Soviet delegate Malik proposes truce in the Korean War.
June 30th – General Ridgway notifies enemy he is ready to discuss possibility of arranging a cease-fire.
July 10th – Truce talks begin at Kaesong; UN delegation led by US Vice-Admiral Charles Turner, Communist group led by Lieutenant-General Nam Il of North Korea.
July 27th – Negotiators at Kaesong agree on agenda.
August 5th – UN Command breaks off truce talks on grounds of armed enemy troops in the neutral area.
August 10th – Cease-fire talks resumed.
August 23rd – Communists suspend cease-fire talks on grounds of 'bombing' of their delegation at Kaesong.
October 25th – Armistice conference resumed at new site, Panmunjom.
October 28th – Agreement reached on battle line as the line of demarcation.
November 12th – Ridgway orders Van Fleet to cease offensive operations and begin active defence of UN front, thus introducing the stalemate which lasted until June, 1952.
November 26th – Agreement reached on location of battle line. 'Little Armistice' begins next day.

December 18th – Prisoner-of-war lists exchanged by both sides.
December 27th – 'Little Armistice' ends but war remains stalemated.

1952

January 2nd – United Nations makes proposal on prisoner exchange embodying the principle of 'voluntary repatriation'.
January 3rd – Communists reject UN proposal in such language as to indicate the 18-month deadlock over voluntary repatriation has been reached.
April 28th – Admiral Joy presents UN final offer insisting on voluntary repatriation.
May 7th – General Mark Clark arrives in Tokyo to succeed Ridgway as Supreme Commander of UN forces.
May 22nd – Major-General William Harrison relieves Admiral Joy as chief of UN delegation at Panmunjom.
October 8th – United Nations adjourns armistice talks indefinitely until Communists accept its proposal on prisoner exchange or make a suitable counter offer.
November 17th – India introduces compromise truce plan at United Nations.
December 15th – Peiping radio announces Communist China's formal rejection of Indian compromise plan.

1953

March 28th – Communists accept UN proposal to discuss exchange of sick and wounded prisoners-of-war.
April 20th – Exchange of sick and wounded prisoners – 'Operation Little Switch' – begun in Korea.
April 26th – Truce talks resumed at Panmunjom.
May 7th – Communists accept a UN proposal that prisoners unwilling to be repatriated be kept in neutral custody in Korea rather than removed to a neutral nation.
May 25th – New proposals for ending prisoner deadlock offered at Panmunjom, but South Korean observer boycotts meetings; beginning of South Korean President Rhee's campaign to block the cease-fire is indicated.
June 8th – Agreement reached on prisoner-of-war issue.
June 9th – South Korean National Assembly unanimously rejects truce terms.
June 14th – Communists launch heaviest offensive in two years at ROK troops in eastern sector.

June 18th – On orders of President Rhee approximately 27,000 North Korean prisoners are freed and returned to civilian life in South Korea.

June 20th – Communists accuse UN Command of complicity in freeing of prisoners; suspend truce talks.

June 23rd – President Rhee reiterates opposition to truce terms.

June 25th – Walter Robertson, US Assistant Secretary of State for Far East, arrives in Seoul to persuade Rhee to accept truce.

July 8th – Communists agree to General Clark's proposal to proceed with final arrangements for armistice without South Korean participation.

July 11th – Robertson and Rhee announce agreement; Rhee will no longer oppose truce terms.

July 13th – Communists launch even larger offensive than June 14th assault against ROK troops.

July 27th – Cease-fire agreement signed, Korean War ends.

August 5th – Exchange of prisoners – 'Operation Big Switch' – begins at Panmunjom.

Appendix B

COMMONWEALTH BATTLE CASUALTIES

August 24th, 1950 to July 27th, 1953

	KILLED IN ACTION		WOUNDED IN ACTION		MISSING*		TOTAL	
	Officers	Other Ranks	Officers	Other Ranks	Officers	Other Ranks	Officers	Other Ranks
Australia ..	9	252	40	994	4	33	53	1,279
Canada ..	11	283	59	1,143	2	45	72	1,471
India ..	—	—	—	4	—	—	—	4
New Zealand	2	20	15	64	1	—	18	84
South Africa	—	—	—	—	1	—	1	—
United Kingdom	71	616	187	2,311	52	1,050	310	3,976
Totals ..	93	1,170	301	4,516	60	1,128	454	6,814
							7,268	

In addition to battle casualties, approximately 11 Officers and 135 Other Ranks died in Korea as a result of accidents or sickness.

* Of the total of 60 Officers and 1,128 Other Ranks shown as missing, 40 Officers and 996 Other Ranks – a total of 1,036 all ranks – were repatriated by the Communists, and one man refused repatriation.

Appendix C

HONOURS AND AWARDS

August 1950 to July 1953

AWARD	AUSTRALIA	CANADA	NEW ZEALAND	UNITED KINGDOM	TOTAL
Victoria Cross	—	—	—	4	4
George Cross	—	—	—	2	2
KBE	—	—	—	1	1
CB	—	1	1	—	2
CBE	1	3	—	4	8
OBE	12	17	—	29	58
MBE	14	58	9	80	161
BEM	2	21	3	30	56
DSO (3rd Bar)	—	—	—	1	1
DSO (2nd Bar)	—	1	—	1	2
DSO (1st Bar)	—	—	—	4	4
DSO	6	8	4	31	49
MC (1st Bar)	1	—	—	6	7
MC	25	33	11	107	176
GM	1	1	—	3	5
DCM (1st Bar)	1	1	—	—	2
DCM	4	7	1	12	24
MM (1st Bar)	1	—	—	—	1
MM	39	52	7	111	209
RRC	—	1	—	—	1
ARRC ..	—	1	—	—	1
Mentioned in Dispatches ..	84	247	45	571	947
Foreign Decorations	17	—	4	77	98

Bibliography

Banner Over Pusan by Ellery Anderson, Evans 1960.
Captive in Korea by Philip Deane, Hamish Hamilton 1953.
Cry Korea by Reginald Thompson, Macdonald 1952.
General Dean's Story as told to William Worden, Weidenfeld & Nicolson 1954.
Korean Reporter by René Cutforth, Brown Watson 1957.
In Spite of Dungeons by Rev. S. J. Davies, Hodder & Stoughton 1954.
Korea and the Fall of MacArthur by Marguerite Higgins, Oxford University Press 1960.
Korean Tales by Melvin B. Voorhees, Secker & Warburg 1953.
No Rice for Rebels by Francis S. Jones, Lane 1956.
Now Thrive the Armourers by Robert Holles, Harrap 1955.
Our Men in Korea by Eric Linklater, HMSO 1952.
South to the Naktong, North to the Yalu by Roy E. Appleman, Office of the Chief of Military History 1960.
Substitute for Victory by John Dille, Doubleday 1954.
The Australians in Korea ed. by Norman Bartlett, Australian War Memorial 1954.
The Dead, the Dying and the Damned by D. J. Hollands, Cassell 1956.
The Edge of the Sword by Anthony Farrar-Hockley, Muller 1954.
The First Commonwealth Division by C. N. Barclay, Gale & Polden 1954.
The Korea Knot by Carl Berger, Oxford University Press 1957.
The Korean War by Robert Leckie, Pall Mall 1963.
The Limited War by David Rees, Macmillan 1964.
The Morning Calm by Hamish Bain, Badger Books 1958.
The War in Korea by Major R. C. W. Thomas, OBE, Gale & Polden 1954.
The Wooden Boxes by Derek Kinne, Muller 1955.
War in Korea by Marguerite Higgins, Oxford University Press.

REGIMENTAL JOURNALS AND BOOKLETS ON KOREA

Canada's Army in Korea.
Princess Patricia's Light Infantry by G. R. Stevens, O.B.E.
St George's Gazette, Journal of the Royal Northumberland Fusiliers.
The Argylls in Korea by G. I. Malcolm.

The Back Badge, Journal of the Gloucestershire Regiment.
The Borderers in Korea by John MacDonald.
The Britannia, Journal of the Royal Norfolk Regiment.
The Crossbelts, Journal of the 8th Hussars.
The Iron Duke, Journal of the Duke of Wellington's Regiment.
The Middlesex Regiment by P. Kemp.
The RedHackle, Journal of the Black Watch.
The Royal Ulster Rifles in Korea by Hugh Hamill.

The Gloucesters by E. J. Kahn, article in the *New Yorker*.

Index

Here are three fascinating books that give an outstanding narrative account of the political history of our century. People, places and events are put into proper perspective. The dazzling progress of the twentieth century is placed under the authors' microscopes for all to see . . .

A HISTORY OF THE WORLD IN THE 20th CENTURY

PART I 1899-1918

D. C. Watt **(50p) 10/-**

Section I: The world on the eve of the twentieth century. Europe at its zenith.
Section II: The breakdown of the European system.
Section III: The war of 1914-1918.

PART II 1918-1945

Frank Spencer **(60p) 12/-**

Section I: The Post War Era.
Section II: From the first New Deal to the Second World War.
Section III: The Second World War.

PART III 1945-1968

Neville Brown **(50p) 10/-**

Section I: From Hiroshima to Panmunjom.
Section II: From Panmunjom to the Moon.